Building Arduino PLCs

The essential techniques you need
to develop Arduino-based PLCs

Pradeeka Seneviratne

Apress®

Building Arduino PLCs: The essential techniques you need to develop Arduino-based PLCs

Pradeeka Seneviratne
Udumulla, Mulleriyawa, Sri Lanka

ISBN-13 (pbk): 978-1-4842-2631-5   ISBN-13 (electronic): 978-1-4842-2632-2
DOI: 10.1007/978-1-4842-2632-2

Library of Congress Control Number: 2017932449

Managing Director: Welmoed Spahr
Editorial Director: Todd Green
Acquisitions Editor: Pramila Balan
Development Editor: Anila Vincent
Technical Reviewer: Jayakarthigeyan Prabakar
Coordinating Editor: Prachi Mehta
Copy Editor: Kezia Endsley
Compositor: SPi Global
Indexer: SPi Global
Artist: SPi Global
Cover image designed by Freepik

Distributed to the book trade worldwide by Springer Science+Business Media New York, 233 Spring Street, 6th Floor, New York, NY 10013. Phone 1-800-SPRINGER, fax (201) 348-4505, e-mail orders-ny@springer-sbm.com, or visit www.springeronline.com. Apress Media, LLC is a California LLC and the sole member (owner) is Springer Science + Business Media Finance Inc (SSBM Finance Inc). SSBM Finance Inc is a **Delaware** corporation.

For information on translations, please e-mail rights@apress.com, or visit http://www.apress.com/rights-permissions.

Apress titles may be purchased in bulk for academic, corporate, or promotional use. eBook versions and licenses are also available for most titles. For more information, reference our Print and eBook Bulk Sales web page at http://www.apress.com/bulk-sales.

Any source code or other supplementary material referenced by the author in this book is available to readers on GitHub via the book's product page, located at www.apress.com/978-1-4842-2631-5. For more detailed information, please visit http://www.apress.com/source-code.

Printed on acid-free paper

Contents at a Glance

Contents

About the Author

Pradeeka Seneviratne is a software engineer with over 10 years of experience in computer programming and systems design. He loves programming embedded systems such as Arduino and Raspberry Pi. Pradeeka started learning about electronics when he was at primary college by reading and testing various electronic projects found in newspapers, magazines, and books.

Pradeeka is currently a full-time software engineer who works with highly scalable technologies. Previously, he worked as a software engineer for several IT infrastructure and technology servicing companies, and he was also a teacher for information technology and Arduino development.

He researches how to make Arduino-based unmanned aerial vehicles and Raspberry Pi-based security cameras.

Pradeeka is also the author of the *Internet of Things* with Arduino Blueprints, Packt Publishing.

About the Technical Reviewer

Jayakarthigeyan Prabakar is an electrical and electronics engineer with more than four years of experience in real-time embedded systems development. He loves building cloud-connected physical computing systems using Arduino, MSP430, Raspberry Pi, BeagleBone Black, Intel Edison, ESP8266, and more.

Jayakarthigeyan started understanding how computing devices and operating systems work when he started repairing his personal computer in middle school. That was when he first got his hands on electronics.

From his third year in the undergraduate degree program, he started building prototypes for various startups around the world as a freelancer. Currently, Jayakarthigeyan is a full-time technical lead of the R&D division in a home automation startup and works as a consultant to many other companies involved in robotics, industrial automation, and other IoT solutions. He helps build prototypes to bring their ideas to reality.

CHAPTER 1

■ ■ ■

Getting Ready for the Development Environment

A Programmable Logic Controller (PLC) is a *digital computer* that continuously monitors or scans the state of input devices and controls the state of output devices based on a custom program. A basic industrial PLC typically consists of an embedded computer, inputs, outputs, and a power supply with battery backup. They usually automate *industrial electromechanical* processes.

Figure 1-1 presents an industrial PLC mounted on a **DIN rail**. This unit consists of separate elements, including a *power supply, controller,* and unit for handling inputs and outputs. Typically for high voltage levels, the input unit consists of optically isolated inputs and output unit consists of *optically isolated* relay outputs. The passive components are enclosures, terminal block connectors, and DIN rails.

Electronic supplementary material The online version of this chapter (doi:10.1007/978-1-4842-2632-2_1) contains supplementary material, which is available to authorized users.

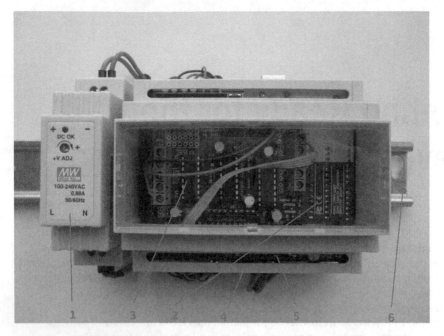

Figure 1-1. *Modules of an Arduino-based PLC*
Image courtesy of Hartmut Wendt at www.hwhardsoft.de

The following are the major components that can be identified in the Figure 1-1.

1. Power supply

2. Controller

3. Relay/non-relay unit for input and output

4. Enclosure

5. Terminal block connectors

6. DIN rail

Arduino Development Environment can be used to build functional PLCs that can be used with some industrial automation and process control. You'll learn how to choose appropriate components for various parts of the PLC, such as the CPU, inputs, outputs, network interfaces, power supplies, and battery backups.

This chapter provides a comprehensive shopping guide to purchasing various assembled printed circuit boards, some of the hardware components (*active and passive*), and setting up your development environment to make all the projects discussed in the chapters in the book.

We'll provide an array of manufacturers and suppliers, but the products may have same core functionalities and slightly different features. A good example is the Arduino UNO board that comes with different features depending on the manufacturer, but uses the same Arduino UNO **bootloader**.

■ **Note**   This guide is only limited to the major hardware components that will be needed to build projects discussed in this book. The information presented here gives you a basic idea when it comes to purchasing those products from various vendors and manufacturers. The detailed technical guide will provide all the information about the products discussed in the respective chapters.

Buying an Arduino

Arduino comes with different flavors, including boards, modules, shields, and kits. The examples and projects discussed in this book use the Arduino UNO board, which is the basic board of the entire Arduino family. There are plenty of Arduino UNO clones and derived boards available and you may be confused about which one to buy. Following are some popular boards that can be used to start building your development environment, and buying one of them is necessary.

Arduino UNO and Genuino UNO

The Arduino online store is a very good way to purchase an Arduino UNO board. Currently, there are two brands available for Arduino. The Arduino UNO is now available for sale (`store-usa.arduino.cc`) in the United States only and the Genuino UNO is available for sale (`store.arduino.cc`) in the rest of the world.

Arduino UNO

You can purchase an Arduino UNO Rev3 board (see Figure 1-2) from the official Arduino store, which is a Dual Inline Package (DIP) type of ATmega328P microcontroller preloaded with Arduino UNO bootloader (it's about $24.95; `http://store-usa.arduino.cc/products/a000066` and `https://www.sparkfun.com/products/11021`).

Figure 1-2. *Arduino UNO Rev3 board. Image courtesy of* `arduino.cc`

Also, the SMD version (Rev3) of this board is also available at the following stores if you'd like to purchase it.

- **Arduino.org**: about €20.90—`http://world.arduino.org/en/arduino/arduino-uno-smd-rev3.html`

- **SparkFun**' about $29.95—`https://www.sparkfun.com/products/11224`

Genuino UNO

Genuino UNO (see Figure 1-3) is identical to the Arduino UNO except the brand name with the same revision that is Rev3. The board is based on the DIP type of ATmega328P microcontroller. (about €20; `https://store.arduino.cc/product/GBX00066`).

Figure 1-3. *Genuino UNO Rev3 board. Image courtesy of arduino.cc*

Cable and Power Supply

Don't forget to buy a USB cable and a power supply to work with the Arduino board.

USB Cable

You can use one of the following **USB cables** or a similar cable to work with Arduino.

- **Adafruit** - USB Cable - Standard A-B - 3 ft/1m (about $3.95; https://www.adafruit.com/products/62)

- **SparkFun** - USB Cable A to B - 6 Foot (about $3.95; https://www.sparkfun.com/products/512)

Power Supply

The Arduino board can be supplied with power between **7-12V** from the DC power jack. Choosing a 9V power supply is sufficient to function the Arduino board properly. Here are some of the power packs that are ready to work with Arduino.

- **Adafruit** -9 VDC 1000mA regulated switching power adapter; UL listed (about $6.95; https://www.adafruit.com/product/63)

- **SparkFun** - Wall Adapter Power Supply - 9VDC 650mA (about $5.95; https://www.sparkfun.com/products/298)

5

Arduino UNO Clones and Derived Boards

There are plenty of Arduino UNO clones and derived boards (also known as *derivatives*) available from various manufacturers. The exact replicas of the Arduino boards with different branding are called clones. Arduino derivatives are different from clones, because they are derived from the Arduino hardware design but provide a different layout and a set of features (i.e., Teensy by PJRC and Flora by Adafruit), often to better serve a specific market. One of the following is a great choice for an alternative Arduino UNO board.

Seeeduino (Figure 1-4) from Seeed Development Limited is a derivative Arduino board that can be used to build Arduino projects instead of using the official Arduino board (about $19.95; https://www.seeedstudio.com/Seeeduino-V4.2-p-2517.html).

Figure 1-4. *Seeeduino v4.2. Image courtesy of Seeed Development Limited*

You will also need a **micro-USB cable** to program this board (about $2.5; https://www.seeedstudio.com/Micro-USB-Cable-48cm-p-1475.html).

SparkFun RedBoard

SparkFun RedBoard (see Figure 1-5) is also a goof solution to use as an alternative Arduino board to build Arduino-based projects (about $19.95; https://www.sparkfun.com/products/12757). This shield brings some favorite features like UNO's optiboot bootloader, the stability of the FTDI, and the R3 shield compatibility.

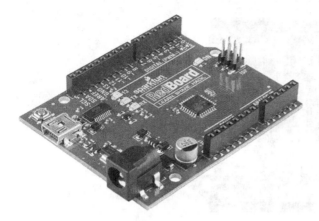

Figure 1-5. *SparkFun RedBoard. Image From SparkFun Electronics; Photo taken by Juan Peña*

You also need a **USB Mini-B cable** to program this board (about $3.95; https://www.sparkfun.com/products/11301). You can power the board over **USB** or through the **barrel jack**.

Buying an Arduino Ethernet Shield

The main functionality of **Arduino Ethernet Shield** is to connect your Arduino board to the Internet. You only need an **Arduino Ethernet Shield** if you are planning to build a **cloud**-connected PLC that will be discussing in **Chapter 8, "Mapping PLCs into the Cloud Using a NearBus Cloud Connector"**.

Arduino Ethernet Shield 2

This is the latest version of the **Arduino Ethernet Shield** (Figure 1-6) manufactured by arduino.org at the time of this writing. It is based on the **Wiznet W5500** Ethernet chip. The shield has a standard **RJ-45** jack, on board **micro-SD card slot**, and six **TinkerKit** connectors. You learn more about Arduino Ethernet in **Chapter 2, "Arduino, Ethernet, and WiFi"** (about €22; http://world.arduino.org/en/arduino-ethernet-shield-2.html).

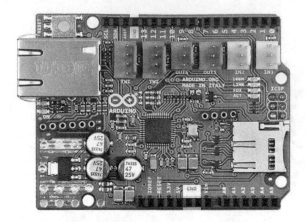

Figure 1-6. Arduino Ethernet Shield 2. Image courtesy of arduino.org

Alternatively, the POE (Power Over Ethernet) version of this board is also available at http://world.arduino.org/en/arduino-ethernet-shield-2-with-poe.html and is about €35.20.

However, you can use the previous version of Arduino Ethernet Shield (Figure 1-7) based on the **Wiznet W5100** Ethernet chip, provided that you already have one and it works well with the projects discussed in this book.

Figure 1-7. Arduino Ethernet Shield (previous version). Image from SparkFun Electronics; photo taken by Juan Peña

Buying an Arduino WiFi Shield

If you'd like to connect your PLC wirelessly to the Internet and build cloud-connected PLCs, this is the best choice.

The Arduino WiFi Shield (Figure 1-8) connects your Arduino board to the Internet wirelessly through WiFi.

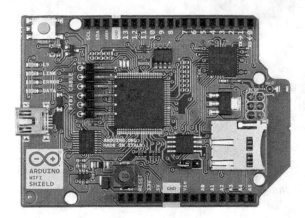

Figure 1-8. *Arduino WiFi shield. Image courtesy of* arduino.org

You will learn more about Arduino WiFi in **Chapter 2,** "Arduino, Ethernet, and WiFi". (about €75.90; http://world.arduino.org/en/arduino/arduino-wifi-shield-antenna-connector.html).

Buying a Grove Base Shield

This is the **Base Shield** (Figure 1-9) we will use for building PLC projects discussed in this book. It is an *Arduino UNO compatible* shield operating with **5V/3.3VDC** directly received from the Arduino board. The shield is easy to use and provides *4-wire* standard *Grove-type connectors* to connect sensors, actuators, and devices, hence no soldering is required and it's easy to plug and play. So this is perfect for prototyping and you can make your prototype neatly without jumper wires. Also, you can quickly plug and remove sensors, actuators, and devices to debug your code.

Figure 1-9. *Grove Base Shield v2.0. Image courtesy of Seeed Development Limited*

Grove provides plenty of sensing and actuating boards with standard 4-wire grove connectors. You simply plug them directly into the shield, to the analog, digital, UART, or I2C female connector.

Grove Base Shield has **three** versions—**v1.1, v1.2, and v2.0**. In this book we'll be using **Grove Base shield v2.0**. However, if you have an older version of the board, you can still keep using it with the projects. The v2.0 shield has 16 grove connectors. In Chapter 4, "Your First Arduino PLC," you learn more about the Grove Base shield.

Buying Grove Components

Grove provides ready-to-use components for sensors and actuators that you can use with Grove Base Shield to quickly set up Arduino projects without using a large amount of wires. The following sections discuss some important Grove components that you will need in order to build Arduino-based PLC projects.

Grove Button

The Grove button (Figure 1-10) is an ideal hardware component to test your PLCs by sending input signals (2-level logic) to Arduino boards through the Grove Base Shield. The Grove button contains a **momentary on/off push button**, **pull-down resistor,** and **standard 4-pin Grove connector**. The push button outputs a HIGH signal when pressed and the LOW signal when released. Get a few of them; they will help you add more inputs (about $1.9; http://www.seeedstudio.com/Grove-Button-p-766.html).

Figure 1-10. *Grove button. Image courtesy of Seeed Development Limited*

Grove LED

Grove LED (Figure 1-11) is an another convenient hardware component that we'll use with projects to *see* the output produced by PLCs. It consists of an LED, brightness controller (potentiometer), and a Grove connector. Get a few of them to use with the projects; they are available in several different colors. (about $1.9; https://www.seeedstudio.com/Grove---Red-LED-p-1142.html).

Figure 1-11. *Grove LED. Image courtesy of Seeed Development Limited*

■ **Note** All Arduino UNO, Arduino UNO clones, and derivative UNO boards such as Seeeduino, RedBoard, and Adafruit have an onboard LED normally connected to the digital pin 13. You can use this LED as a simulation of output.

Grove Relay

Grove Relay (Figure 1-12) can be used to drive a high load from the Arduino board. The board consists of a Normally Open relay, LED indicator, standard Grove connector, and a few electronic components. The peak voltage capability is 250VAC at 10amps (about $2.9; https://www.seeedstudio.com/Grove---Relay-p-769.html).

Figure 1-12. *Grove Relay. Image courtesy of Seeed Development Limited*

Grove Temperature Sensor

The **Grove Temperature Sensor** (Figure 1-13) can be used to measure ambient temperature in the range of -40 to 125 °C with an accuracy of 1.5°C. It outputs variable voltages depending on the temperature that is turned by the on-board voltage divider (about $2.9; https://www.seeedstudio.com/Grove-Temperature-Sensor-p-774.html).

Figure 1-13. *Grove Temperature Sensor. Image courtesy of Seeed Development Limited*

Grove Speaker

Grove Speaker (Figure 1-14) is another output device that you can use with PLCs to make outputs audible. The board equipped with a small speaker, volume control, standard Grove connector, and a few electronic components (about $6.9; https://www.seeedstudio.com/Grove---Speaker-p-1445.html).

Figure 1-14. Grove Speaker. Image courtesy of Seeed Development Limited

Grove Infrared Reflective Sensor

Object detection is helpful for ensuring the presence of an object or set of objects and for generating output signals accordingly. In industrial process automation, these sensors play a major role in actuating different mechanical devices and making them start functioning properly. For example, you could use an infrared reflective sensor to detect the presence of a bottle in the production line and actuating a label passing device.

Grove Infrared Reflective Sensor (Figure 1-15) is an ideal solution to quickly set up as an object detection sensor with Arduino-based PLCs. This board consists of an **IR LED** and a **photosensor pair**. The sensor produces **digital HIGH** when the reflected light is detected. If no reflection detected, it produces **digital LOW**. It comes with a standard Grove interface that can be directly plugged in to the Grove Base Shield (about $4.9; https://www.seeedstudio.com/Grove-Infrared-Reflective-Sensor-p-1230.html).

Figure 1-15. Grove Infrared Reflective Sensor. Image courtesy of Seeed Development Limited

Grove Cables

Don't forget to buy a few more **Grove cables** (Figure 1-16) to connect your inputs and outputs to the Grove Base Shield. The connector is universal since you can plug it to either the **analog, digital, UART**, or **I2C** *connector* on the **Grove Base Shield**. Grove cables come with different lengths and types. The lengths are **5cm, 20cm, 30cm, 40cm, and 50cm**. Most of them are buckled and a few are unbuckled. Each cable consists of four wires—red, black, white, and yellow.

Figure 1-16. *Grove Universal 4-Pin Buckled Cable. Image courtesy of Seeed Development Limited*

Buying a Relay Shield

Relay plays a major role in PLCs to latch the output signals. There are various Arduino UNO compatible relay shields available, but we'll present two relay shields that can be easily used for working with the projects. They can be easily seated on the Arduino UNO with wire wrap headers without soldering, hence they are easy to plug and remove. These relay shields can be used to build applications that implement multiple relay outputs. Typically they will provide four outputs or more.

Arduino 4 Relays Shield

The **Arduino 4 Relays Shield** (Figure 1-17) allows you to drive high-power loads that are rated with high current and **voltages up to 48VDC;** Arduino can't directly power them through the digital pins.

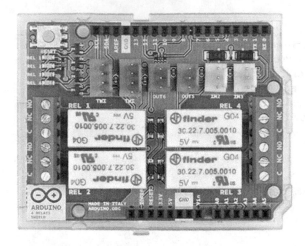

Figure 1-17. Arduino 4 Relays Shield. Image courtesy of arduino.org

The shield can only handle **four output devices** and it has **two TinkerKit inputs, two TinkerKit outputs,** and **two TinkerKit TMI interfaces**. You'll learn in-depth about this relay shield in Chapter 9, "Building a Better PLC" (about €22; http://world. arduino.org/en/arduino-4-relays-shield.html).

SeeedStudio Relay Shield

Same as the Arduino 4 Relays Shield, the **SeeedStudio Relay Shield** (Figure 1-18) also allows you to drive **high-power loads** that are rated with high current and voltages up to 35VDC, 120VAC or 250VAC, which Arduino can't directly power through the digital pins.

Figure 1-18. SeeedStudio Relay Shield. Image courtesy of Seeed Development Limited

You'll learn in-depth about this relay shield in Chapter 9, "Building a Better PLC" (about $20; https://www.seeedstudio.com/Relay-Shield-v30-p-2440.html).

Buying an ArduiBox

ArduiBox (Figure 1-19) is a DIY kit for **Arduino UNO**, **Arduino 101**, and **Arduino Zero**. It allows you to install your Arduino-based PLC in a control cabinet and mount it to a **DIN rail** like any other industrial PLC available in the market.

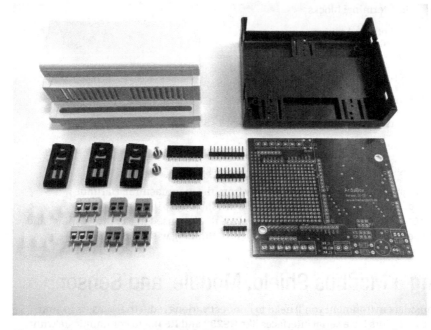

Figure 1-19. *Components of an ArduiBox. Image courtesy of Hartmut Wendt* *www.hwhardsoft.de*

■ **Note** *DIN* stands for Deutsches Institut fur Normung (German Institute of Standardization), which specifies a metal rail of a standard type for mounting circuit breakers and industrial control equipment inside equipment racks. It's known as a DIN rail. Typically, DIN rails are made out of cold rolled carbon steel sheet with a zinc-plated or chrome-plated bright surface finish. Visit www.din.de/en for more information about the German Institute of Standardization.

At the time of this writing, the **ArduiBox kit** was available for €34.99 including optional parts. The kit doesn't include any Arduino or shield.

The kit includes:

- Milled DIN rail enclosure with transparent top
- pcb with prototyping board and landing zone for Arduino and shield
- 2x 3-terminal blocks
- 4x 2-terminal blocks
- Sockets for Arduino
- Sockets for shield
- 2x self-tapping screws

And has following optional parts:

- Parts for 12V voltage regulator (Vin 15 - 30VDC)
- Reset button
- Additional terminal block for voltage regulator

You can purchase a ArduiBox kit directly from the manufacturer, Hartmut Wendt (http://www.hwhardsoft.de/english/webshop/raspibox/#cc-m-product-10145780397) if you'd like to build a PLC that's housed in a control cabinet and mount it in a DIN rail as you'll learn in Chapter 5, "Building with ArduiBox".

Buying a Modbus Shield, Module, and Sensor

In an industrial environment you'll need to connect various industrial sensors to your PLCs. These sensors have serial interfaces like **RS232** and **RS485** to communicate with computers using the Modbus communications protocol.

To enable your Arduino PLC with the Modbus communications protocol, you'll have to prepare your toolbox with the following boards. Note that they are a bit expensive.

In Chapter 7, "Modbus," you'll learn how to connect an industrial temperature sensor to your Arduino PLC, read temperature values, and make decisions accordingly.

Multiprotocol Radio Shield for Arduino

The Multiprotocol Radio Shield (Figure 1-20) from CookingHacks is an Arduino UNO compatible shield that's ideal for building Modbus-enabled PLCs. The shield is designed to connect two communication modules at the same time.

Figure 1-20. *Multiprotocol Radio Shield for Arduino. Image courtesy of* www.cooking-hacks.com

(It's about €33.00; https://www.cooking-hacks.com/multiprotocol-radio-shield-board-for-arduino-rpi-intel-galileo.)

RS485/Modbus Module for Arduino and Raspberry Pi

The **RS485 Module for Arduino and Raspberry Pi** (Figure 1-21) allows you to connect more than one industrial devices to Arduino with only two wires.

Figure 1-21. *RS485/Modbus Module for Arduino. Image courtesy of* www.cooking-hacks.com. *TQS3-I: MODBUS RS485 Interior Thermometer*

(About €35; https://www.cooking-hacks.com/rs-485-modbus-module-shield-board-for-arduino-raspberry-pi-intel-galileo.)

The **TQS3-I - MODBUS RS485 Interior Thermometer** (Figure 1-22) from PAPOUCH (www.papouch.com) supports **Modbus** and Spinel communication protocols via an RS485 bus line. It can be used with Multiprotocol Radio Shield and RS485 Modbus modules for Arduino to easily set up temperature sensor projects (about $57; www.papouch.com).

Figure 1-22. *TQS3-I: MODBUS RS485 Interior Thermometer. Image courtesy of* www.*papouch*.com

Downloading Software

All the code examples in this book have been implemented and tested on a Windows operating system environment. You'll need the following software to download and save them to your local drive.

Generally, the following compressed files will be saved by default in your Windows computer's downloads folder. You will learn how to install, run, and configure them in the next chapters, when required.

Arduino Software

Download the latest **Arduino Software** from https://www.arduino.cc/en/Main/ Software (Figure 1-23). Use the **Windows zip file for non-admins. You run it** directly from your folder after it's been copied to the local drive without installing it on your computer. You can also download it directly from https://www.arduino.cc/download_ handler.php?f=/arduino-1.6.11-windows.zip.

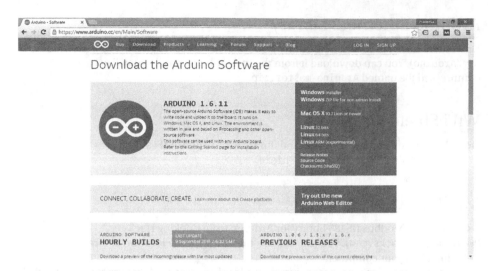

Figure 1-23. Arduino Software Download page

plcLib

You can download the plcLib library for Arduino from http://www.electronics-micros.com/resources/arduino/plclib/plcLib.zip or from GitHub's master branch at https://github.com/wditch/plcLib. It contains the latest commits of the library (Figure 1-24).

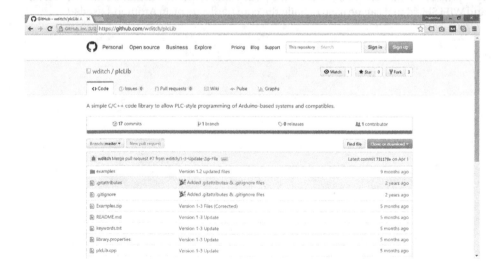

Figure 1-24. plcLib download page at GitHub

Arduino Ethernet2 Library

Download the Arduino Ethenet2 library from GitHub (https://github.com/arduino-org/Arduino). You can download it from the master branch and you will get a compressed file named Arduino-master.zip.

WiFi Shield Firmware

Download the up-to-date firmware and library for **ArduinoWiFi shield** at https://github.com/arduino/wifishield.

Modbus RS485 Library

Download the Modbus RS485 library for Arduino from Cooking Hacks (www.cooking-hacks.com). You can also use the following file location to directly download it to your computer:
     https://www.cooking-hacks.com/media/cooking/images/documentation/RS-485/RS485_for_Arduino.zip

Summary

In this chapter you learned about how to prepare your development environment with ready-to-use hardware and software components. You gained a basic understanding of the important hardware and software components that can be used to build Arduino-based PLC projects. In next chapter, you learn the basics about Arduino, Arduino Ethernet Shield, and WiFi Shield that can be used to build the core hardware of an Arduino-based PLC.

CHAPTER 2

■ ■ ■

Arduino, Ethernet, and WiFi

Arduino can work as the heart of the PLCs that we are going to build. The basic type of Arduino, **Arduino UNO,** is best for learning *basic concepts* behind the PLCs as well as implementing prototypes with a combination of **hardware** (inputs, outputs) and **software** (Arduino software and libraries like plcLib and Modbus).

In the previous chapter, you learned about how to prepare your development environment with various hardware and software. In this chapter, you will learn more about **Arduino UNO**, **Arduino Ethernet shield,** and **Arduino WiFi shield**.

Arduino and Genuino

The **Arduino UNO** and **Genuino** boards are identical and share the same quality of manufacturing. The boards are based on the **ATmega328P microcontroller** that is powered with **Arduino UNO bootloader**. The microcontroller has **32KB** of **flash memory** and **2KB** of **RAM**. Arduino boards are best for learning electronics and for rapid prototyping. You can interface your Arduino board with various sensors and actuators. At the time of this writing, the **revision** number of Arduino was **3** (Rev3).

Figure 2-1 presents the main components of the Arduino UNO/Genuino board.

Figure 2-1. *Arduino UNO Rev3 board*

© Pradeeka Seneviratne 2017
P. Seneviratne, *Building Arduino PLCs*, DOI 10.1007/978-1-4842-2632-2_2

The main components are marked as follows:

1. ATmega328P microcontroller

2. USB jack (type B)

3. Power jack (9V wall wart or 9V battery)

4. Digital pins

5. Analog pins

6. Power IN (9V external)

7. Power OUT

8. Power indicator

9. LED connected to digital pin 13

10. Voltage regulator

11. Reset button

Digital Pins

Digital pins can be used to interface with various **sensors** and **actuators;** Arduino UNO has **14 digital pins**. These pins can be configured as either input or output with Arduino software. All digital pins operate at **5V** and they can receive or produce about **20mA** of current, but make sure not to supply them with more than **40mA** of current to avoid permanent damage to the microcontroller. The *default state* of a digital pin is **input** and each digital pin has an internal pull-up resistor (disconnected by default) of 20-50kilo ohm by default.

Some digital pins have specialized functions:

- *Transmit and receive serial data:* Digital pins 0 and 1 can be used to perform serial communications, where pin 0 can receive (RX) and pin 1 can transmit (TX) TTL serial data, respectively.

- *Pulse width modulation:* Digital pins 3, 5, 6, 9, 10, and 11 can be used to provide 8-bit PWM output.

- *External interrupts:* Digital pins 2 and 3 can be configured to trigger an interrupt on a low value, a rising or falling edge, or a change in value. These pins are very useful when you want to make alternative paths in your control process using switches or sensors, and counting pulses from sensors.

Analog Pins

The **Arduino UNO** has **six analog pins** that can be used to input analog signals; they are labeled starting from **A0** to **A5**. You can read analog-to-digital converted values from **0 to 255,** which represents the range **0V to 5V**. These pins are very useful when you want to read values from analog sensors, like *temperatures, humidity, proximity,* and many more.

Powering the Arduino Board

Arduino boards can be powered with different power sources and use relevant methods, depending on the nature of your project.

USB Power

You can connect the Arduino UNO board to a computer with a **USB Type A/B cable** (Figure 2-2). Just connect the **Type B** end of the cable to **Arduino's USB jack** and the **Type A** end to the computer's **USB port.** The Arduino board can use USB power while uploading sketches and testing with a computer or preforming serial communications with a computer. Alternatively, you can use any type of **5V USB power supply** if you are not planning to use the USB cable as the communications link between Arduino and the computer.

Figure 2-2. *Arduino UNO powered with USB*

9V AC/DC Adapter

This is also called a **wall wart** (Figure 2-3). However, you can use an **AC/DC adapter** that can supply between **7-12V** with a **center-positive connector**.

Figure 2-3. *9V wall wart. Image From SparkFun Electronics; Photo taken by Juan Peña*

9V Battery Pack

You can power your Arduino UNO board with a **9V battery** attached to a **9V battery holder** (Figure 2-4) with a center-positive barrel jack.

Figure 2-4. *9V battery with a holder. Image From SparkFun Electronics; photo taken by Juan Peña*

VIN Pin

You can use a **VIN pin** to supply between **7-12V DC** without using a DC plug (Figure 2-5). The VIN pin is not polarity protected so make sure to connect the positive lead of the power supply to it.

Figure 2-5. *Arduino UNO powered through a VIN pin with an external 9V power source. Image credited to the "original creator"*

The power source is automatically either a USB or an external source, depending on the *input voltage* by the *on-board* **LMV358 OP-AMP**.

Arduino Ethernet

The Arduino Ethernet shield connects your Arduino board to the Internet and it can be used to build interactive projects for the Internet of Things (IoT) in conjunction with Arduino software and the Arduino Ethernet library.

Arduino Ethernet Shield 2

The **Arduino Ethernet shield 2** is based on a **WIZnet W5500** Ethernet chip with on-board *RJ-45 jack, Micro-SD card slot, and TinkerKit connectors*. This shield is manufactured by arduino.org but this is not the latest version of Arduino Ethernet shield manufactured by arduino.cc. You can use the **Arduino Ethernet 2 library** to write sketches for the Arduino Ethernet Shield.

Figure 2-6 presents some important components on the Arduino Ethernet Shield 2.

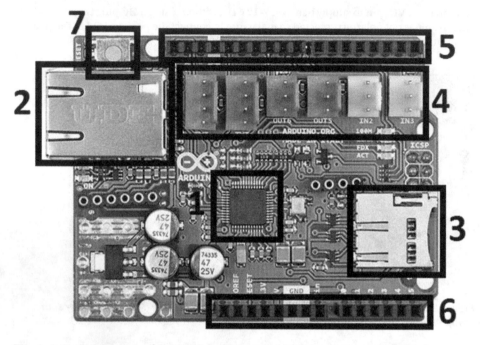

Figure 2-6. *Components of the Arduino Ethernet Shield 2. Image courtesy of* `arduino.org`

The main components are marked as follows:

1. WIZnet W5500 Ethernet Chip

2. RJ45 jack

3. Micro-SD slot

4. TinkerKit connectors

5. Headers (digital and PWM)

6. Headers (analog and power)

7. Reset button

Table 2-1 lists the technical specifications of the **Arduino Ethernet Shield 2** published by `arduino.org`.

Table 2-1. *Technical Specifications of the Arduino Ethernet Shield 2*

Component	Specification
Chip Based	WIZnet W5500
Card Reader	Micro-SD
Ethernet	802.3 10/100 Mbit/s
ThinkerKit Interface	2x TWI, 2x OUT, 2x IN
Interfaces with Arduino Board	SPI
Input Voltage Plug (limits)	6-20V
Input Voltage POE (limits)	36-57V
Operating Voltage	5V
Current Needs	87ma (with webserver sketch and 9V on the DC plug)

Source http://www.arduino.org/products/shields/arduino-ethernet-shield-2

Connecting Them Together

To connect an Arduino Ethernet Shield 2 to the Internet, you need the following:

- RJ-45 cable (Figure 2-7)

Figure 2-7. *Cat 6 Ethernet cable. Image from SparkFun Electronics; photo taken by Juan Peña*

- Router or switch (Figure 2-8)

  An Internet connection is required only if you are planning to access your board through the Internet by port forwarding or a dedicated IP.

Figure 2-8. *Router with four Ethernet ports. Source* `https://en.wikipedia.org/wiki/Router_(computing)#/media/File:Adsl_connections.jpg`

Now connect the Arduino Ethernet Shield to the Arduino board using its **wire-wrap headers** (Figure 2-9).

Figure 2-9. *Arduino Ethernet Shield 2 wire-wrap headers. Image courtesy of* `arduino.org`

Make sure to properly seat it on the Arduino board. The Ethernet board draws power from the Arduino board and shares the same pin layout through the headers. The **Reset button** on Ethernet Shield 2 can be used to reset the Arduino.

You can still use the Arduino Ethernet Shield Rev3 (Figure 2-10) that is manufactured by arduino.cc but it seems to be discontinued. Search the Internet for ***Arduino Ethernet Shield*** to find online stores if it's available. If you already have one, use the *Arduino Ethernet library* instead of *the Arduino Ethernet2 library*. Note that the Arduino Ethernet Shield Rev3 is not the *predecessor* to the Arduino Ethernet Shield 2.

Figure 2-10. *Arduino Ethernet Shield Rev3. Image courtesy of* arduino.cc

MAC Address

When you write a sketch for the Arduino Ethernet shield, you should provide a unique **MAC (Media Access Control)** address for it. A MAC address is a globally unique identifier for a network device. If you don't know the dedicated MAC address of your Ethernet shield, you can use a random MAC address as long as it does not conflict with the network devices in your local network.

Usually the MAC address of the Ethernet shield can be found on the **bottom of the shield** (Figure 2-11). However, you can still use your own MAC address and override the factory assigned address.

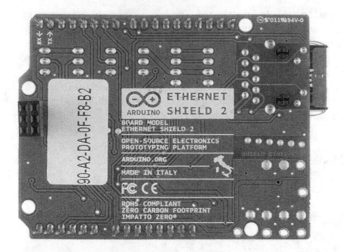

Figure 2-11. *MAC address printed on the sticker. Image From SparkFun Electronics; photo taken by Juan Peña*

When you are writing sketches for the Ethernet shield, the MAC address should be rewritten **in hexadecimal** format with prefixes of 0x and comma-separated. The 0x prefix indicates the number is in hexadecimal rather than in some other base. The MAC address printed on the sticker in Figure 2-11, **90-A2-DA-0F-F8-B2,** can be written as **0x90, 0XA2, 0xDA, 0x0F, 0XF8, 0XB2** in hexadecimal.

IP Address

In addition to the MAC address, the Ethernet shield should be configured with an **IP address**. You can assign a static IP address within the sketch or, if not, the router will dynamically assign an IP address to the Ethernet shield.

If you want to assign a static IP address, the IP address should be in a comma-separated format by replacing all the dots (dotted-decimal notation) with commas. For an example, the IP address **192.168.0.177** can be written as **192, 168, 0, 177** in your Arduino sketch.

Arduino WiFi

Instead of a wired connection to the Internet, Arduino can also connect to the Internet with a **wireless** connection. The **Arduino WiFi shield** (Figure 2-12**)** provides wireless connection through a WiFi network to the Internet, and you can build the projects by replacing the Ethernet shield with a WiFi shield. You can use Arduino with a WiFi shield for handheld and portable projects. Also, it avoids having to work with tedious network cables and long-range supports and saves you time.

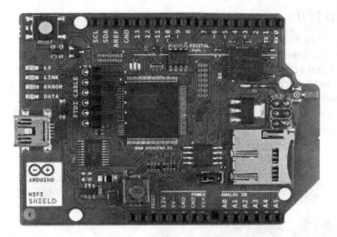

Figure 2-12. *The Arduino WiFi Shield, top view. Image courtesy of arduino.org*

The Arduino WiFi shield can be connected to the Arduino UNO board using the soldered *wire-wrap headers*, same as the Ethernet shield. It provides the same pin layout through the headers and draws power from Arduino UNO board, and it shares the Reset button to reset the complete stack.

A very good reference about the WiFi shield can be found at http://www.arduino. org/learning/getting-started/getting-started-with-arduino-wifi-shield.

Arduino Software

Arduino software is an **Integrated Development Environment (IDE)** that allows you to write **sketches** (program) and upload them to an Arduino board. The Arduino development environment is written in Java and based on **Processing** (processing.org) and other open source software.

Downloading Arduino Software

Using the instructions provided in the previous chapter, you can download the Arduino Windows ZIP file for non-admin install to your Windows computer. The advantage of this is you don't need to install it on your computer and you can run it by just double-clicking on the executable file.

However, you can download and try the Windows installer instead and it will work the same as the non-admin installation.

Using the Arduino IDE

Now extract the downloaded ZIP file (i.e., `arduino-1.6.11-windows.zip`) using the **WinZip** or **7-Zip** compression software.

You will get a folder named `arduino-1.6.11`. Inside the folder you will find an executable file named `arduino`, which is an application type file. Simply double-click on the icon to start the Arduino IDE.

If it prompts you with a Windows Security Alert. Just click on the Allow Access button to proceed.

The IDE is loaded with a default sketch (Figure 2-13). The menu bar and toolbar provide a host of options and configurations to work with Arduino boards.

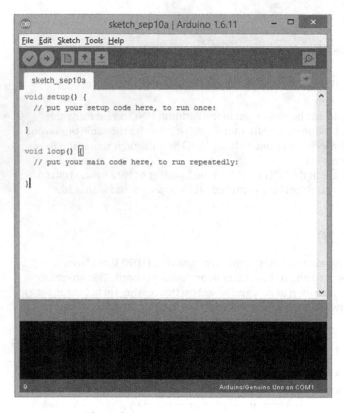

Figure 2-13. *The Arduino IDE*

Where Is the libraries Folder?

All libraries that can be used to write Arduino sketches are located inside the libraries folder. Generally, it is a top-level folder (Figure 2-14).

```
..\arduino-1.6.11-windows\arduino-1.6.11\libraries
```

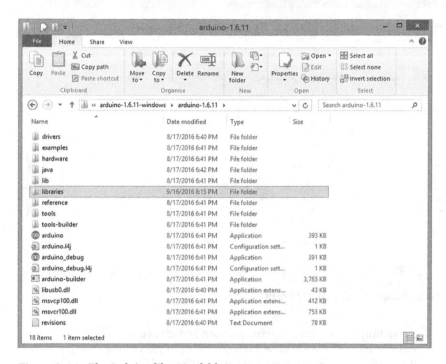

Figure 2-14. *The Arduino libraries folder*

Adding the Ethernet2 Library

The **Arduino Ethernet2 library** isn't included with Arduino IDE developed and provided by arduino.cc. Therefore, you should manually add it to your Arduino IDE's libraries folder.

1. Download the Arduino Ethernet2 library from **GitHub** (https://github.com/arduino-org/Arduino). You can download it from the master branch and you will get a file in zip format. To download the file as a zip file, click **Clone or download** button in the top-right corner of the browser window, then form the drop-down menu, click **Download ZIP** link.

35

2. Extract the zip file and navigate to the libraries folder. You can see a folder named Ethernet2 (Figure 2-15). Now copy the Ethernet2 folder to Arduino IDE's libraries folder. Finally, restart the Arduino IDE.

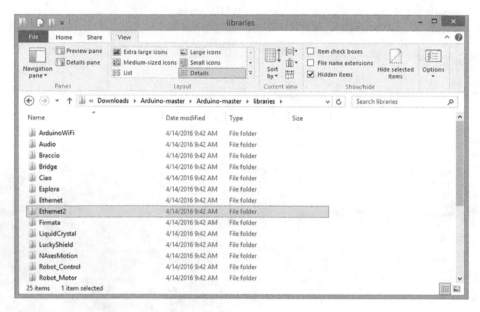

Figure 2-15. *Finding the Ethernet2 library*

3. To confirm the Ethernet2 library has been added to your Arduino IDE, choose Sketch ➤ Include Library. If you can see the menu item called Ethernet2 under Contributed Libraries, you are ready to go.

Cables

The communication link between the Arduino board and your computer can be established using a **USB type A/B cable** (See Chapter 1, "Getting Ready for the Development Environment," for a shopping guide).

The USB type A/B cable consists of USB type A and type B female connectors. Connect the type A connector side to your computer's USB port and the type B connector side to your Arduino's USB port (Figure 2-16). The Arduino board can use the DC power supplied by the USB port from your computer, which is typically 5V DC.

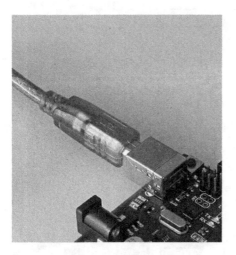

Figure 2-16. *USB type A/B cable (type B end) connected to Arduino UNO. Image credited to the "original creator"*

Basic Configurations

Before uploading any sketches to your Arduino board, you should configure the following components correctly.

1. *Board Type:* This is the board you have currently attached to your computer. On the menu bar, select the correct board type by choosing Tools ➤ Board ➤ Arduino/Genuino UNO.

2. *COM Port:* This is the computer's COM port, which is what your Arduino board is currently attached to. You can find the COM port number under the Device Manager in Windows (Figure 2-17).

Figure 2-17. *Identifying the COM port assigned to Arduino*

Now, on the menu bar, choose Tools ➤ Port and then select the correct port from the list.

Writing Sketches for Arduino UNO

You have set up your development environment with Arduino software, as well as with various hardware including boards, shields, USB cables, and power supplies. Now we are going to take a look at how to write sketches for Arduino, mainly focusing on building PLCs.

Bare Minimum Code

An Arduino *code* or *program* is generally called a **sketch**. The minimum sketch that needs to be compiled on Arduino software is called **the** bare minimum code. The sketch consists of two functions—setup() and loop()—as shown in Listing 2-1.

Listing 2-1. Bare Minimum Code Example (BareMinimum.ino)

```
void setup() {
  // put your setup code here, to run once:

}
```

```
void loop() {
  // put your main code here, to run repeatedly:

}
```

The setup() function is called when a sketch starts and the loop() function is called after that. The loop() function loops consecutively whatever things are mentioned inside it, until you press the **Reset** button or until the next power cycle.

When you start the Arduino IDE or create a new file, it opens a default sketch file with the bare minimum code.

Now, let's upload the bare minimum code into the Arduino board. Remember to configure the Arduino IDE with your board, such as board type and COM port, before proceeding.

1. Save the file in to your computer's local drive by choosing **File ➤ Save As** and typing the filename as BareMinimum. Click the **Save** button to save the file.

   Click the **Verify** button to compile the code. You'll get output on the **Messages** section of the IDE that's similar to the following (Figure 2-18).

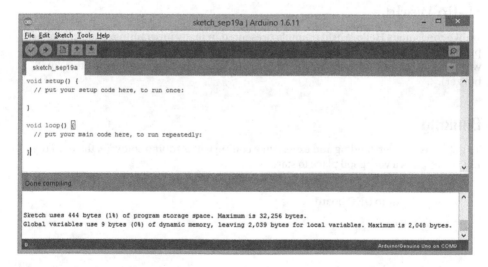

Figure 2-18. *Uploading a sketch from Arduino IDE*

2. Click **Upload** button to upload the code to the Arduino board. While uploading the sketch to the board, you can see the orange color **RX** and **TX** LEDs on the board are flashing (Figure 2-19). The RX LED blinks when the Arduino is receiving data and the TX LED blinks when the Arduino is transmitting data.

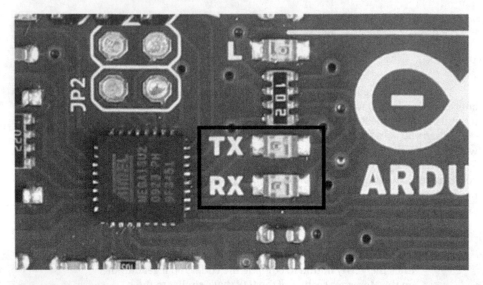

Figure 2-19. Arduino UNO's RX and TX LEDs

Hello World

Unlike the famous **Hello World** example presented in many programming books, which prints the text **Hello World** on a console, Arduino can provide the next level of greeting with physical output—that is, making an LED blink (or you can make beeps with a small buzzer).

Blinking

To get a basic understanding and experience coding with Arduino software, the blinking LED example is a very good place to start.

You'll need following components to build the hardware setup:

- Arduino UNO board

- LED

- 220 Ohm resistor

- Small breadboard

Now connect the LED with the Arduino board, as shown in Figure 2-20. You can use a small breadboard to hook up the circuit using hook-up wires. The following steps explain how to build it:

1. Connect the short leg of the LED to the Arduino GND.

2. Connect the long leg of the LED to the Arduino digital pin 13 through a 220 ohm resistor. The resistor works as a current-limiting resistor to limit the LED's current.

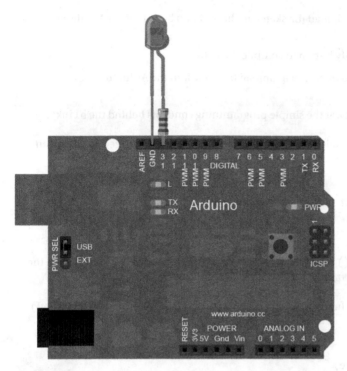

Figure 2-20. *LED blink circuit*

Now you need an Arduino sketch to blink the LED. A simple sketch can be found in the Arduino examples folder. To open the sketch, choose File ➤ Examples ➤ Basic ➤ Blink under Built-in Examples from the Arduino IDE's menu bar. A new window will open with the following sketch (Listing 2-2).

Listing 2-2. LED Blink Example (Blink.ino)

```
// the setup function runs once when you press reset or power the board
void setup() {
  // initialize digital pin 13 as an output.
  pinMode(13, OUTPUT);
}

// the loop function runs over and over again forever
void loop() {
  digitalWrite(13, HIGH);   // turn the LED on (HIGH is the voltage level)
  delay(1000);                       // wait for a second
  digitalWrite(13, LOW);    // turn the LED off by making the voltage LOW
  delay(1000);                       // wait for a second
}
```

Now you are ready to upload the sketch to the Arduino board. Using Arduino IDE, do the following.

1. Click **the Verify** button to compile the code.

2. Click **the Upload** button to upload the code into the Arduino board.

Let's briefly take a look at the simple programming concepts behind the blink sketch.

Inside the setup() function, the Arduino digital pin 13 has initialized as an **output** pin by passing the OUTPUT constant as a parameter to the pinMode() function.

```
void setup() {
  // initialize digital pin 13 as an output.
  pinMode(13, OUTPUT);
}
```

Also, inside the loop() function, the digitalWrite() function is used to change the voltage level on a particular Arduino digital pin. The voltage level should be either 5V (HIGH) or 0V (LOW).

The digitalWrite() function accepts values for duty cycles between 0 (always off) and 255 (always on).

Syntax:

```
digitalWrite(pin, value)
```

Parameters:

pin: The pin number to write to
value: HIGH or LOW

Returns:

None

Now, to supply 5V to the Arduino digital pin 13, you can write the code as follows:

```
digitalWrite(13, HIGH);
```

Likewise, to supply 0V to the Arduino digital pin 13, you can write the code as follows:

```
digitalWrite(13, LOW);
```

To make the LED blink, you should add time delays between the **HIGH** and **LOW** voltage states. The delay() function can be used to make delays and it accepts the delay time in milliseconds.

In this example, the delay between **on** and **off** is one second. You can make different blinking effects by changing the delay time.

The digitalWrite() function is really useful in PLC development projects to output digital signals.

Fading

Now you may have a question about how to work with more than two voltage levels on an Arduino pin. This can be done using Arduino analog pins. Let's create a quick circuit by removing the **220 ohm resistor** leg from the **digital pin 13** and connect it to the **PWM pin 9**. Figure 2-21 presents the circuit built on a breadboard.

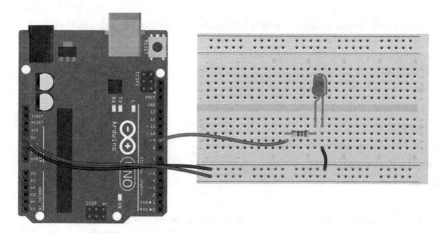

Figure 2-21. *LED fading circuit*

Now open Arduino IDE and choose File ➤ Examples ➤ Basics ➤ Fade to open the example sketch (Listing 2-3).

Listing 2-3. LED Fading Example (Fade.ino)

```
int led = 9;           // the PWM pin the LED is attached to
int brightness = 0;    // how bright the LED is
int fadeAmount = 5;    // how many points to fade the LED by

// the setup routine runs once when you press reset:
void setup() {
  // declare pin 9 to be an output:
  pinMode(led, OUTPUT);
}

// the loop routine runs over and over again forever:
void loop() {
  // set the brightness of pin 9:
  analogWrite(led, brightness);

  // change the brightness for next time through the loop:
  brightness = brightness + fadeAmount;
```

```
  // reverse the direction of the fading at the ends of the fade:
  if (brightness <= 0 || brightness >= 255) {
    fadeAmount = -fadeAmount;
  }
  // wait for 30 milliseconds to see the dimming effect
  delay(30);
}
```

Let's learn some programming concepts behind this code that is related to the analog aspect of Arduino.

The `analogWrite()` function accepts values for duty cycles between 0 (always off) and 255 (always on).

Syntax:

analogWrite(pin, value)

Parameters:

`pin`: The PWM pin number to write to

`value`: The duty cycle (0-255)

Returns:

None

Initially, the brightness value is set to 0 and then it increases by 5 using the `fadeamount` variable. When it becomes greater than or equal to 255 or less than or equal to 0, the initial value of `fadeamount` gets inverted and starts to reverse the direction of the fading.

```
  // reverse the direction of the fading at the ends of the fade:
  if (brightness <= 0 || brightness >= 255) {
    fadeAmount = -fadeAmount;
  }
```

The delay between each fading state is set to 30 milliseconds using the `delay()` function.

Reading Analog Inputs

The Arduino board has specialized pins for reading analog signals or data from various sensors and potentiometers. In this simple project, you learn how to read analog data from a potentiometer connected to the Arduino **analog pin A0**. To build this circuit, you need a 10kilo ohm potentiometer.

Build the circuit as shown in Figure 2-22.

1. Connect the center pin of the potentiometer to the Arduino analog pin A0.

2. Connect one of the outer pins to the Arduino 5V pin.

3. Connect the other (remaining) outer pin to the Arduino GND.

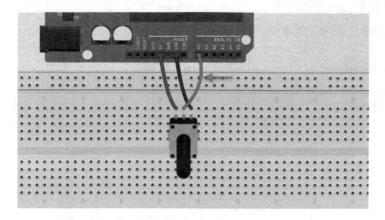

Figure 2-22. *Analog read circuit*

4. Now open the sample sketch by choosing File ➤ Examples
    ➤ Basics ➤ AnalogReadSerial. The sample sketch will open
    in a new window. Verify the sketch and upload it to the
    Arduino board. Follow the analog read serial example shown
    in Listing 2-4.

Listing 2-4. Analog Read Serial Example (AnalogReadSerial.ino)

```
// the setup routine runs once when you press reset:
void setup() {
  // initialize serial communication at 9600 bits per second:
  Serial.begin(9600);
}

// the loop routine runs over and over again forever:
void loop() {
  // read the input on analog pin 0:
  int sensorValue = analogRead(A0);
  // print out the value you read:
  Serial.println(sensorValue);
  delay(1);        // delay in between reads for stability
}
```

5. Open the **Serial Monitor** by choosing Tools ➤ Serial Monitor.
    The Serial Monitor will open in a new window (Figure 2-23)
    and display the current value of the analog pin A0. The output
    will continuously scroll until you disable the **Autoscroll**
    option by deselecting the checkbox. The output value is
    between 0 to 1023, so it is a fraction of 5V. Actually, you can
    calculate the current voltage on the analog pin by using this
    value. Use the following formula to calculate the voltage.

■ **Note**   Voltage = value of the analog read x (5/1023)

As an example, the voltage for an analog read of 500 would be:

```
500 x (5/1023) = 2.44V
```

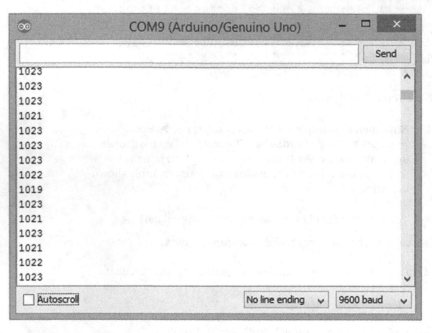

Figure 2-23. *Arduino Serial Monitor*

Another tool you can use to see the output is the **Serial Plotter, which is found on the Tools** menu (Figure 2-24). It provides a graphical representation of the output.

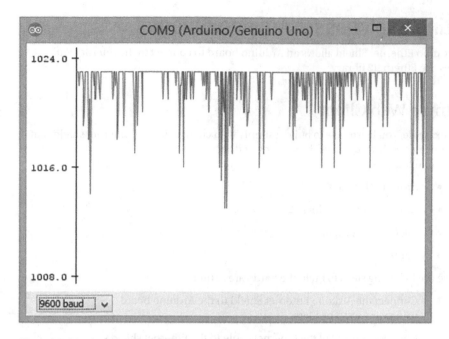

Figure 2-24. *Arduino Serial Plotter*

Inside the setup() function, the serial communication between the Arduino and the computer is initialized to 9600 bits per second using Serial.begin().

```
Serial.begin(9600);
```

The begin() function of the Serial class accepts the baud rate in bits per second. It opens the serial port for communication and sets the data rate to **9600bps**. 9600 bits per second is the default for the Arduino, and is perfectly adequate for the majority of users, but you could change it to one of these rates: 300, 600, 1200, 2400, 4800, 9600, 14400, 19200, 28800, 38400, 57600, or 115200.

The sensor value on the analog pin can be read with the analogRead() function and it accepts the analog pin number as the parameter.

```
int sensorValue = analogRead(A0);
```

Syntax:

```
analogRead(pin)
```

Parameters:

pin: Analog pin number A0 to A5

Returns:

Integer value (0 to 1023)

47

Writing Sketches for Arduino Ethernet

The Arduino Ethernet Shield allows an Arduino board to connect to the Internet and requires the Ethernet2 library.

A Simple Web Client

In this example, you learn how to build a simple web client using the Ethernet shield that communicates with a web server by sending and receiving data.

To build this project, you need following things to configure the hardware setup:

- Arduino UNO board

- Arduino Ethernet Shield 2

- Cat 6 Ethernet cable

- Router

Use the following steps to build the hardware setup:

1. Connect the Arduino Ethernet Shield to the Arduino board using wire-wrap headers.

2. Connect one end of the Ethernet cable to the Ethernet shield's Ethernet jack and the remaining end to the router or switch.

3. Connect the Arduino UNO board to the computer with a USB type A/B cable.

Now you are ready to upload the sample sketch provided with the Arduino IDE.

1. Open the Arduino IDE and on the menu bar and choose File ➤ Examples ➤ Ethernet 2 ➤ WebClient. A sample sketch for a web client (Listing 2-5) will open in a new window.

Listing 2-5. Web Client Example (WebClient.ino)

```
#include <SPI.h>
#include <Ethernet.h>

// Enter a MAC address for your controller below.
// Newer Ethernet shields have a MAC address printed on a sticker on the
shield
byte mac[] = { 0xDE, 0xAD, 0xBE, 0xEF, 0xFE, 0xED };
// if you don't want to use DNS (and reduce your sketch size)
// use the numeric IP instead of the name for the server:
//IPAddress server(74,125,232,128);  // numeric IP for Google (no DNS)
char server[] = "www.google.com";    // name address for Google (using DNS)
```

```
// Set the static IP address to use if the DHCP fails to assign
IPAddress ip(192, 168, 0, 177);

// Initialize the Ethernet client library
// with the IP address and port of the server
// that you want to connect to (port 80 is default for HTTP):
EthernetClient client;

void setup() {
  // Open serial communications and wait for port to open:
  Serial.begin(9600);
  while (!Serial) {
    ; // wait for serial port to connect. Needed for native USB port only
  }

  // start the Ethernet connection:
  if (Ethernet.begin(mac) == 0) {
    Serial.println("Failed to configure Ethernet using DHCP");
    // try to configure using IP address instead of DHCP:
    Ethernet.begin(mac, ip);
  }
  // give the Ethernet shield a second to initialize:
  delay(1000);
  Serial.println("connecting...");

  // if you get a connection, report back via serial:
  if (client.connect(server, 80)) {
    Serial.println("connected");
    // Make a HTTP request:
    client.println("GET /search?q=arduino HTTP/1.1");
    client.println("Host: www.google.com");
    client.println("Connection: close");
    client.println();
  } else {
    // if you didn't get a connection to the server:
    Serial.println("connection failed");
  }
}

void loop() {
  // if there are incoming bytes available
  // from the server, read them and print them:
  if (client.available()) {
    char c = client.read();
    Serial.print(c);
  }
```

```
// if the server's disconnected, stop the client:
if (!client.connected()) {
  Serial.println();
  Serial.println("disconnecting.");
  client.stop();

  // do nothing forevermore:
  while (true);
  }
}
```

2. Change the MAC address according to your Ethernet shield or just keep it as it is.

```
byte mac[] = { 0xDE, 0xAD, 0xBE, 0xEF, 0xFE, 0xED };
```

3. Next, change the IP address. Be sure to use one within your network's valid IP address range.

```
IPAddress ip(192, 168, 0, 177);
```

4. Make sure your router has an active Internet connection, because you are going to access a live URL (www.google.com).

5. Click the **Verify** button to compile the sketch. You'll probably get the following warning regarding the newly added Ethernet2 library because the properties file of the library doesn't have a category mentioned.

```
Warning Category '' in library Ethernet2 is not valid. Setting to
'Uncategorized'
```

A quick fix can be applied to resolve this issue by adding an entry to the library. properties file in the libraries/Ethernet2 folder. Add the following line just after the paragraph= entry and then save the file (Figure 2-25).

```
category=Communication
```

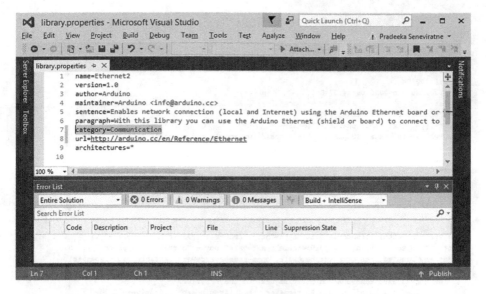

Figure 2-25. *Add the proper category to the library file*

Now try to verify the sketch again. You'll notice that the warning about the missing category is now resolved.

6. Click the **Upload** button to upload the sketch to the Arduino board. This will take a few seconds to complete the process.

7. Now open the **Serial Monitor** by choosing Tools ➤ Serial Monitor (Ctrl+Shift+M). You'll see the following output on the Serial Monitor. Note that the section between <html></html> contains HTML output from google.com for the search query arduino (Figure 2-26).

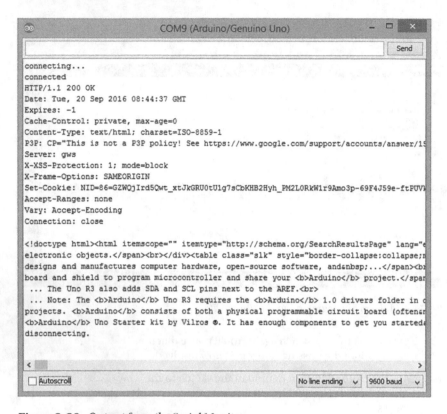

Figure 2-26. *Output from the Serial Monitor*

This is only a basic example that shows how to use the Ethernet Shield to communicate with the Internet.

Writing Sketches for Arduino WiFi

To write sketches for the Arduino WiFi Shield, you need to use the Arduino WiFi library.
The WiFi shield can connect to three types of WiFi networks:

- **Open networks:** Requires only the SSID of the network.

- **WPA networks:** Requires the SSID and password of the network.

- **WPE networks:** Requires the SSID and key of the network.

To get started with the WiFi shield, let's upload a simple sketch to the Arduino board that can be used to connect it to a WPA network.

1. Choose File ➤ Examples ➤ WiFi ➤ ConnectWithWPA. The ConnectWithWPA example sketch (Listing 2-6) will open in a new window.

Listing 2-6. WiFi Connection with WPA Example (ConnectWithWPA.ino)

```
#include <SPI.h>
#include <WiFi.h>

char ssid[] = "yourNetwork";      //  your network SSID (name)
char pass[] = "secretPassword";   // your network password
int status = WL_IDLE_STATUS;      // the Wifi radio's status

void setup() {
  //Initialize serial and wait for port to open:
  Serial.begin(9600);
  while (!Serial) {
    ; // wait for serial port to connect. Needed for native USB port only
  }

  // check for the presence of the shield:
  if (WiFi.status() == WL_NO_SHIELD) {
    Serial.println("WiFi shield not present");
    // don't continue:
    while (true);
  }

  String fv = WiFi.firmwareVersion();
  if (fv != "1.1.0") {
    Serial.println("Please upgrade the firmware");
  }

  // attempt to connect to Wifi network:
  while (status != WL_CONNECTED) {
    Serial.print("Attempting to connect to WPA SSID: ");
    Serial.println(ssid);
    // Connect to WPA/WPA2 network:
    status = WiFi.begin(ssid, pass);

    // wait 10 seconds for connection:
    delay(10000);
  }

  // you're connected now, so print out the data:
  Serial.print("You're connected to the network");
  printCurrentNet();
  printWifiData();

}
```

```
void loop() {
  // check the network connection once every 10 seconds:
  delay(10000);
  printCurrentNet();
}

void printWifiData() {
  // print your WiFi shield's IP address:
  IPAddress ip = WiFi.localIP();
  Serial.print("IP Address: ");
  Serial.println(ip);
  Serial.println(ip);

  // print your MAC address:
  byte mac[6];
  WiFi.macAddress(mac);
  Serial.print("MAC address: ");
  Serial.print(mac[5], HEX);
  Serial.print(":");
  Serial.print(mac[4], HEX);
  Serial.print(":");
  Serial.print(mac[3], HEX);
  Serial.print(":");
  Serial.print(mac[2], HEX);
  Serial.print(":");
  Serial.print(mac[1], HEX);
  Serial.print(":");
  Serial.println(mac[0], HEX);

}

void printCurrentNet() {
  // print the SSID of the network you're attached to:
  Serial.print("SSID: ");
  Serial.println(WiFi.SSID());

  // print the MAC address of the router you're attached to:
  byte bssid[6];
  WiFi.BSSID(bssid);
  Serial.print("BSSID: ");
  Serial.print(bssid[5], HEX);
  Serial.print(":");
  Serial.print(bssid[4], HEX);
  Serial.print(":");
  Serial.print(bssid[3], HEX);
  Serial.print(":");
  Serial.print(bssid[2], HEX);
  Serial.print(":");
```

```
Serial.print(bssid[1], HEX);
Serial.print(":");
Serial.println(bssid[0], HEX);

// print the received signal strength:
long rssi = WiFi.RSSI();
Serial.print("signal strength (RSSI):");
Serial.println(rssi);

// print the encryption type:
byte encryption = WiFi.encryptionType();
Serial.print("Encryption Type:");
Serial.println(encryption, HEX);
Serial.println();
}
```

2. Replace "yourNetwork" with the actual network SSID.

   ```
   char ssid[] = "yourNetwork";
   ```

3. Replace "secretPassword" with the network password.

   ```
   char pass[] = "secretPassword";
   ```

4. Now verify and upload the sketch to the Arduino board.

5. Open the **Serial Monitor** by choosing Tools -> Serial Monitor.

6. The **Serial Monitor** will output details about the status of the network connection and its technical details.

The Arduino WiFi library provides classes and functions that can be used to work with different types of WiFi networks. The following are some good references about the Arduino WiFi library.

- `http://www.arduino.org/learning/getting-started/getting-started-with-arduino-wifi-shield`

- `https://www.arduino.cc/en/Reference/WiFi`

Summary

In this chapter, you learned the basics about how to build simple circuits with Arduino, Arduino Ethernet Shield, and Arduino WiFi Shield by using the Arduino software and basic electronic components. In next chapter, you will learn about some industrial PLCs built with Arduino as the core hardware.

CHAPTER 3

■ ■ ■

Arduino at Heart

A simple *bottle filling* process control system is a good example to understand the fundamentals of PLCs. *An assembly line consists of a conveyor belt that's connected to a motor drive, nozzle, and a filling stroke. The conveyer belt moves to one direction with empty bottles on it and the presence of a bottle is captured by a sensor at the point of the nozzle. It stops the conveyor belt, starts the liquid pump to fill the bottle, detects the liquid level of the bottle by another sensor, stops the liquid pump, and starts the conveyor belt again. The process continues again and again until the process control system receives an interrupt.*

This process can be controlled by a piece of software that is running on an **embedded** computer. Figure 3-1 presents the graphical representation of a bottle filling system that is controlled by a PLC. Note that the **conveyor**, **nozzle**, and **filling stroke** are connected to the PLC with a set of **sensors** and **actuators**.

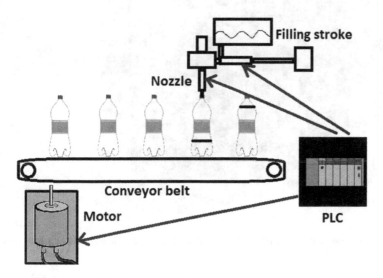

Figure 3-1. *Bottle filling control using PLC*

The software scans all the inputs and process them according to the *user-defined set of instructions, makes decisions,* and *produces outputs* for another set of *user-provided devices.*

© Pradeeka Seneviratne 2017

P. Seneviratne, *Building Arduino PLCs*, DOI 10.1007/978-1-4842-2632-2_3

In this example, the presence of a bottle at the nozzle can be detected using an *infrared sensor* or *proximity sensor*. The *liquid level* inside the bottle can also be detected using an *infrared sensor*. These sensors can produce **inputs** to the PLC. The **actuators** (**output devices**) can be controlled using digital or analog **output** signals coming from the PLC. For this process control, the *conveyor* can be controlled using a *relay controlled motor*, the *nozzle* can be controlled using a *set of servo motors*, and the *filling stroke* can be controlled using an *electric valve*.

What Is PLC?

PLC is a solid state industrial control device that receives inputs from devices such as sensors and switches, processes them using a piece of software with user-defined instructions and logics, and provides outputs for devices such as relays and motors to control the entire process.

Instead of using a *general-purpose computer* to control industrial manufacturing processes, PLCs build flexible, modular based, and easy-to-program automation systems.

Figure 3-2 shows a basic PLC that's mounted on a DIN rail with a power supply unit.

Figure 3-2. *A basic PLC. Image courtesy of Hartmut Wendt Hard & Softwareentwicklung*

The *heart* of a PLC is a **microcontroller** that's capable of running the user-provided software. The *opto-isolated* inputs and outputs can be used to protect the microcontroller unit from external circuitry. These opto-isolated inputs and outputs are exposed through the terminal blocks to connect user-provided devices such as sensors, switches, and actuators. A power supply unit can be mounted on the same DIN rail to use with the PLC to supply regulated voltage and adequate current. Some PLCs have built-in or additional communication modules such as Ethernet or WiFi to connect to the Internet. Most of the PLCs have a USB interface to program the microcontroller's flash memory by using a computer.

Basically, a PLC consists of the following parts and accessories.

- Microcontroller unit (CPU)

- I/O module

- Digital inputs/outputs

- Analog inputs

- PWM outputs

- Power supply unit

- Enclosure

- DIN rail

There are much more advanced PLCs for use in industrial process automation, but we'll only discuss the basic PLCs throughout this book.

■ **Note**   The National Electrical Manufacturers Association (NEMA) defines a PLC as a "digitally operating electronic apparatus that uses a programmable memory for the internal storage of instructions by implementing specific functions, such as logic, sequencing, timing, counting, and arithmetic to control through digital or analog I/O modules various types of machines or processes."

Arduino at Heart

There are some PLCs *based on* **Arduino** and currently available in the market for industrial use. In this section, you learn about the flavors and features of Arduino-based PLCs from various manufacturers.

Industruino

Industruino (https://industruino.com/) is based on an **Arduino Leonardo** *compatible* board, which is another flavor of board in the Arduino family.

Industruino comes with two different flavors:

- Industruino Proto Kit

- Industruino Industrial I/O Kit

Industruino Proto Kit

The **Industruino Proto Kit** (Figure 3-3) consists of the following accessories:

- Topboard (AT32U4 or AT90USB1286 microcontroller and LCD screen)

- Proto baseboard

- Mounting hardware

- Enclosure with membrane button panel

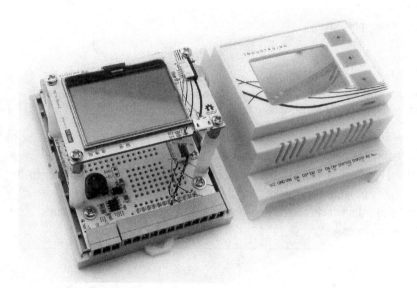

Figure 3-3. *Industruino Proto Kit. Image courtesy of Industruino (https://industruino.com/)*

The **topboard,** which is the *heart* of Industruino, can be purchased based on two types of microcontrollers. They are:

- ATmega 32u4 (32KB flash; Leonardo compatible)

- ATmega AT90USB1286 (128KB flash)

The **ATmega AT90USB1286** microcontroller-based topboard provides **128KB** of *flash memory* to execute large Arduino sketches. The microcontroller unit is soldered on the backside of the PCB.

The topside of the topboard consists of the following components:

- UC1701-based 128x64 LCD

- FPC connector for three-button membrane panel connected

- Micro-USB connector for programming

Protoboard offers a large *prototyping* area to build small electronic circuits that interface with Arduino. It also provides a **14-pin IDC expansion port** and a **5V/2A out voltage regulator** that can be used with your prototypes.

Industruino Industrial I/O Kit

The **Industrial I/O Kit** (Figure 3-4) can be used to build industrial-level automation systems. The Industrial I/O Kit includes the following parts:

- Topboard (AT32U4 or AT90USB1286 microcontroller and LCD screen)

- IND-I/O baseboard (Industrial I/O interface board)

- Mounting hardware

- Enclosure with membrane button panel

Figure 3-4. *Industruino Industrial I/O Kit. Image courtesy of Industruino (https://industruino.com/)*

The **Topboard** is identical to the **Proto Kit** and the **Baseboard** offers a host of interface options with industrial *peripherals*.

Industrial Shields

Industrial Shields comes with a basic controller and two ranges of PLCs, which are based on the Arduino platform. They are:

- 20 I/Os

- Ethernet PLC

Each product has different number of *I/O ports, input voltages,* and *output voltages,* so you can select a suitable product according to your project requirements.

All PLCs are equipped with an original Arduino board with industry grade enclosures that can be mounted on a DIN rail. All PLCs can be programmed using the Arduino IDE and support **USB**, **serial**, **RS232**, **RS485,** and *I2C communications protocols.*

At the time of writing this book, the **20 I/Os PLC range** consisted of two PLC products. They were:

- PLC Arduino ARDBOX 20 I/Os Analog 7.0

- PLC Arduino ARDBOX PLC 20 I/Os RELAY 7.0

Figure 3-5 shows the **PLC Arduino ARDBOX 20 I/Os Analog 7.0.**

Figure 3-5. *PLC Arduino ARDBOX 20 I/Os Analog 7.0. Image courtesy of Industrial Shields (http://www.industrialshields.com/)*

Figure 3-6 shows the **PLC Arduino ARDBOX PLC 20 I/Os RELAY 7.0**.

Figure 3-6. *PLC Arduino ARDBOX PLC 20 I/Os RELAY 7.0. Image courtesy of Industrial Shields (http://www.industrialshields.com/)*

In addition to that, the **Ethernet PLCs** provide Ethernet connectivity to connect the PLC to the Internet and can be used to build industrial **IoT** (**Internet of Things**) projects. Figure 3-7 shows one of the Ethernet PLC products mentioned as **M-DUINO PLC Arduino 19R I/OS Relay/Analog/Digital.**

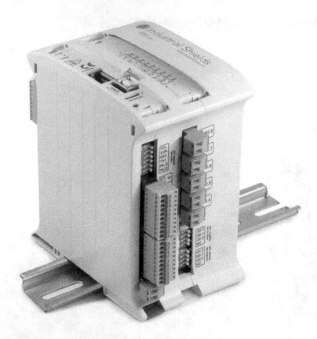

Figure 3-7. *M-DUINO PLC Arduino 19R I/OS Relay/Analog/Digital. Image courtesy of Industrial Shields (*http://www.industrialshields.com/*)*

Presenting complete technical specifications of the **Industrial Shields** is out of the scope of this book. You can learn more about Industrial Shields products by visiting Industrial Shield's official web site at http://www.industrialshields.com/. Also, Industrial Shields provides useful accessories such as *12V DC power supplies, 24V DC power supplies, D-SUB 37 wires, Industrial pull-up I2C connections, and different types of USB cables* that can be used with the range of PLCs. They can be found at http://www.industrialshields.com/open-source/plc-accessories/.

Controllino

Controllino is **Arduino** *compatible* and known as the first software **open-source PLC**. There are three types of **Controllino** PLCs currently available in the market, all with different specifications.

Controllino MINI

The **Controllino MINI** (Figure 3-8) is the basic model of the Controllino family and is based on **Atmel ATmega328** and is similar to the **Arduino UNO**.

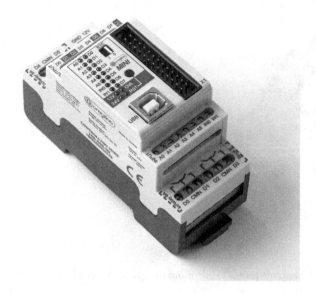

Figure 3-8. *Controllino MINI. Image courtesy of Controllino (http://controllino.biz/)*

Specifications:

- Atmel clock speed: 16MHz

- RTC

- 1x serial interface

- 1x SPI interface

- 1x I2C interface

- Input current max. 8A

- 6x relays outputs: – 230V/6A

- 8x analog/digital inputs

- 8x digital outputs: 2A @12V or 24V

Controllino MAXI

The **Controllino MAXI** is an (Figure 3-9) advance model that's based on **Atmel ATmega2560** and is similar to the **Arduino MEGA**.

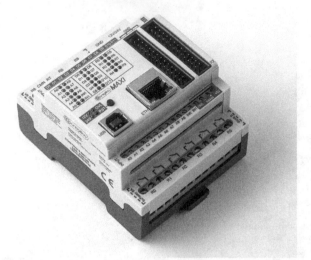

Figure 3-9. *Controllino MAXI. Image courtesy of Controllino (http://controllino.biz/)*

Specifications:

- Clock speed: 16MHz

- RTC

- Ethernet connector

- 2x serial interface

- 1x RS485 interface

- 1x I2C interface

- 1x SPI interface

- Input current Max. 20A

- 10x relays outputs: 230V/6A

- 12x analog/digital inputs

- 12x digital outputs: – 2A @12V or 24V

Controllino MEGA

The **Controllino MEGA** (Figure 3-10) is the most advanced model. It's based on **Atmel ATmega2560** and is similar to the **Arduino MEGA**.

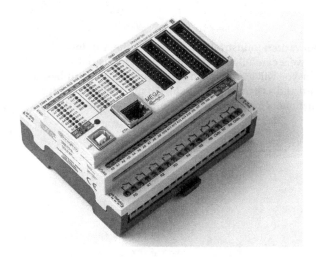

Figure 3-10. *Controllino MEGA. Image courtesy of Controllino (http://controllino.biz/)*

Specifications:

- Clock speed: 16MHz

- RTC

- Ethernet connector

- 2x serial interface

- 1x RS485 interface

- 1x I2C interface

- 1x SPI interface

- Input current Max. 30A

- 16x relays outputs: – 230V/6A

- 21x analog/digital inputs

- 12x digital outputs – high side switch – 2A @12V or 24V

- 12x digital outputs – half-bridge – 2A @12V or 24V

You can learn more about **Controllino** products, their architecture, and installation details at http://controllino.biz/.

Summary

In this chapter, you learned about various industrial PLCs based on the Arduino development environment. They were carefully developed to use in industrial environments. In the next chapter, you will learn how to build your first Arduino-based PLC using an **Arduino UNO** board and a **Grove Base Shield**. Finally, you'll learn how to connect various sensors and actuators to your basic PLC and write simple Arduino sketches for it.

CHAPTER 4

■ ■ ■

Your First Arduino PLC

In the previous chapters, you learned a lot about the fundamentals of PLCs and their applications in industrial process automations. Further, you learned about some Arduino-based industrial PLCs available in the market, which exposed core functionalities of the Arduino with additional hardware layers.

Grove Base Shield Basics

Grove Base Shield is the same size as an **Arduino UNO** board and can be used to create another level of hardware interface with Arduino pins. As you can see in Figure 4-1, there are only three active components soldered on to the board (two resistors and LEDs) and all other components are passive. It has the same pinout as the Arduino UNO, which can be accessed through the on-board wire-wrap headers. Also, these wire-wrap headers can be used to connect the Grove Base Shield to the Arduino board. Figure 4-1 shows the *front view* of the **Grove Base Shield V2**.

Figure 4-1. *Grove Base Shield v2, front. Image courtesy of Seeed Development Limited*

© Pradeeka Seneviratne 2017

P. Seneviratne, *Building Arduino PLCs*, DOI 10.1007/978-1-4842-2632-2_4

On the backside of the shield, you can see that the wire-wrap headers are marked with a similar Arduino UNO pin layout. Figure 4-2 shows the back view of the **Grove Base Shield V2**.

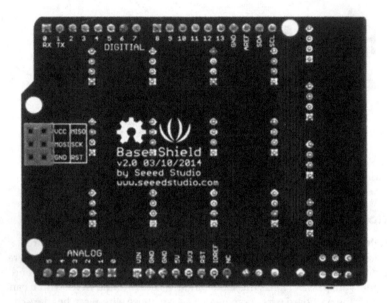

Figure 4-2. *Grove Base Shield v2, back. Image courtesy of Seeed Development Limited*

If you are planning to use a **Grove Base Shield V2** with **Seeeduino V3**, solder the pads **P1** and **P2** (Figure 4-3).

Figure 4-3. *P1 and P2 solder pads. Image courtesy of Seeed Development Limited*

Power Switch

A power switch can be used to select the correct power from the Arduino board. The selectable voltage levels are **3.3V** or **5V**. When you are using the *Grove Base Shield* with *Arduino UNO,* the power switch should be in the 5V position. However, some microcontroller boards, like **Seeeduino Arch**, operate only at **3.3V**. Therefore, select the power depending on the supply voltage of the microcontroller board you are going to use (Figure 4-4).

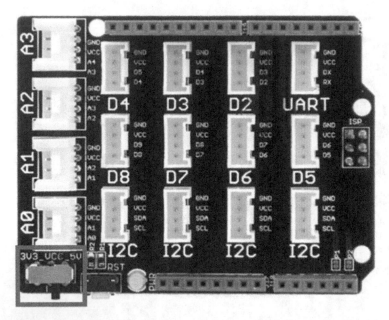

Figure 4-4. Power switch. Image courtesy of Seeed Development Limited

Power Indicator

The **green LED** that reads **PWR** indicates the presence of power (Figure 4-5). It can be either **3.3V** or **5V,** depending on the supply power of the base microcontroller board.

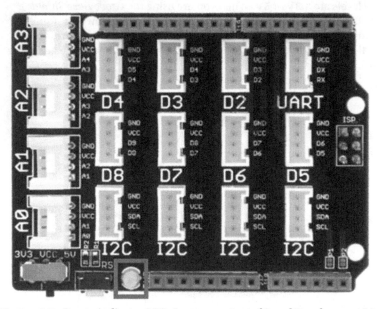

Figure 4-5. Power indicator LED. Image courtesy of Seeed Development Limited

Reset Button

This is a small *momentary push button*, which is connected parallel to the *Arduino reset button* (Figure 4-6). You can reset the Arduino by pressing any reset button. Later in this chapter, we provide instructions for adding your own reset button to a Grove Base Shield.

Figure 4-6. *Reset button. Image courtesy of Seeed Development Limited*

Grove Connectors

The **Grove Base Shield** has **16 Grove connectors** soldered on to the PCB and each connector has four pins. The Grove connector exposes a standard interface for Grove devices, such as **Grove Button, Grove LED, Grove Speaker**, **Grove Temperature Sensor**, and many more.

All *Grove connectors* are physically identical (same dimensions with four pins), but are specialized for different purposes, such as *analog, digital, UART, and I2C*. Table 4-1 shows the specifications for each Grove connector.

Table 4-1. *Specifications for Grove Ports*

Specification	Grove Connector(s)	Qty
Analog	A0,A1,A2,A3	4
Digital	D2,D3,D4,D5,D6,D7,D8	7
UART	UART	1
I2C	I2C	4

73

Digital Ports

A **Grove Digital Port** consists of four pins —**GND**, **VCC**, and *two* **digital** pins. Each Grove digital port is labeled with the pin number of the *outer digital pin,* which is adjacent to the plastic wall of the connector. Figure 4-7 represents the Grove digital port number 2, which is **D2**.

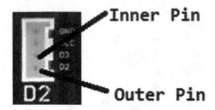

Figure 4-7. Grove digital port D2

Connecting a Grove device to the **D2** port would result the same as if it were connected to the *Arduino digital pin 2.* Table 4-2 shows you the relationship between the *Grove ports* and the *Arduino digital pins.*

Table 4-2. Digital Pin Mapping Between Grove Digital Ports and Arduino Digital Pins

Grove Digital Port	Mapped with Arduino Digital Pin
D2	2
D3	3
D4	4
D5	5
D6	6
D7	7
D8	8

Also, you can connect any digital input or output device to a Grove port by simply using *hookup wires.* As an example, you can connect an LED to the **Grove port D2** by using the pins **D2** and **GND** as a simple hack.

Analog Ports

A **Grove Analog Port** is physically identical to the Grove digital port and consists of four pins—**GND**, **VCC**, and *two* **analog** pins. A *Grove Analog port* is labeled with the pin number of the *outer analog pin,* which is adjacent to the plastic wall of the connector. The analog pin mapping between the *Grove Analog* ports and the *Arduino analog pins* is shown in Table 4-3.

Table 4-3. Analog Pin Mapping Between Grove Analog Ports and Arduino Analog Pins

Grove Analog Port	Mapped with Arduino Analog Pin
A0	0
A1	1
A2	2
A3	3

UART Port

The **UART** (Universal Asynchronous Receiver/Transmitter) **port** consists of four pins—**GND**, **VCC**, **DX**, and **RX**. This port can be used with the devices that are capable of performing serial communications. The **DX** and **TX** pins are internally mapped to the *Arduino* **TX (digital 0)** and **RX (digital 1),** respectively. *Grove Serial LCD* and *Grove UART WiFi* are some of the UART-based devices that can be connected to using the UART port.

I2C Ports

I2C ports allow you to communicate with **I2C/TWI devices**. There are *four I2C* ports and they all share the same pins—**GND**, **VCC**, **SDA**, and **SCL**. The **SDA** (*data*) and **SCL** (*clock*) pins are internally connected to the *Arduino SDA (analog A4)* and *SCL (analog A5)* pins (Figure 4-8). The *Grove I2C Touch sensor, I2C LCD*, and *I2C Motor Driver* are some devices that can be plugged into the *Grove I2C* ports.

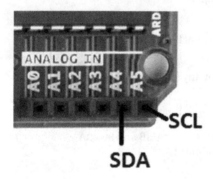

Figure 4-8. Arduino UNO SDA and SCL

■ **Note** The I2C protocol is originally invented and introduced by **Philips**. I2C is a serial protocol for a two-wire interface to connect low-speed devices. These devices can be microcontrollers, EEPROMs, analog-to-digital converters, digital-to-analog converters, and I/O interfaces. However, TWI is used by manufacturers like **Atmel** to refer to their I2C interface to avoid trademark conflicts with Philips, since I2C is a registered trademark. Some manufacturers have implemented proprietary features on top of I2C.

Building a Basic Programmable Logic Controller

In this chapter, you learn how to build a basic Arduino-based PLC step-by-step. This **PLC** operates on **5V DC** power that is supplied by the power supply. The *onboard voltage regulator* (Figure 4-9) regulates the supplied voltage between **7-12V DC** to **5V DC**. The onboard voltage regulator is rated for a maximum of **1000mA**.

Figure 4-9. *Arduino UNO on-board voltage regulator*

Therefore, Arduino can power only small loads attached to it, but basically there are a number of limiting factors.

- The absolute maximum power rating for any single IO pin is 40mA.

- Total current form of all the IO pins together is 200mA.

- The 5V pin can supply current up to 400mA on a USB and 900mA with an external power supply rated to about 1A.

- The 3.3V pin can supply current up to 150mA.

A current limiting resistor would be helpful to limit the amount of current that flows through the load. A good example is to use a **220 Ohm resistor** in series with an **LED** that can drive an *Arduino IO pin* safely.

If you connect a load that can draw high current, the Arduino on-board regulator will heat up, and when it gets overheated, at some point it will automatically shut down temporarily.

The Requirements and Logic

Assume you need to control a simple process using with a **momentary push button** and an actuator. According to the state of the *momentary push button,* the PLC should produce a signal on a particular output line to actuate a device (*LED*).

Generally, the ideal state of a *momentary push button* is **LOW** or **open circuit**. So this state is suitable for **0V** *input*. When you press the *momentary push button*, the circuit becomes *closed* and it can provide a **5V** signal to the input line.

The *momentary push button* positions related to the LED status are shown in Table 4-4.

Table 4-4. *Momentary Push Button Positions and LED Status*

Button Position	LED
IDEAL	OFF
PRESSED	ON
RELEASED	OFF

A piece of **embedded software** (i.e. *an Arduino software-based program*) continuously scans the input line and produces either 0V or 5V on the output line according to the logic provided by the user. The actuator will turn on or off, depending on the voltage it receives.

Required Hardware

To build this project, you need the following hardware.

- Arduino UNO
- Grove Base Shield
- Grove button
- Grove LED
- Grove speaker
- 9V DC power supply
- USB type A/B cable

Connecting the Components

The following steps describe how to connect all the hardware components together.

1. The **Grove Base Shield** comes with *wire-wrap headers* that are soldered to the board, similar to other *Arduino shields*. Just connect the Grove Base Shield to the Arduino UNO board using wire-wrap headers. Make sure to seat it properly on top of the Arduino board.

2. Slide the *power switch* VCC to **5V**.

3. Connect the **Grove Button** to the *connector* marked with **D2** (*Grove port number 2*) using a **Grove cable**. The *connector* **D2** internally connects to *Arduino digital pin 2*.

4. Connect the **Grove LED** to the *connector* marked with **D3**. The *connector* **D3** (*Grove port number 3*) internally connects to *Arduino digital pin 3*.

Figure 4-10 shows the completed hardware setup for the PLC.

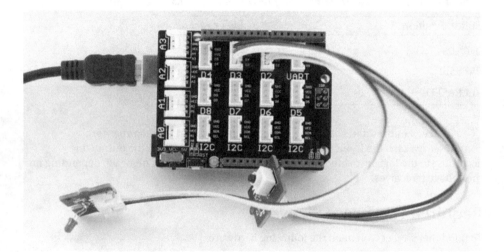

Figure 4-10. *Hardware setup for the PLC. Image courtesy of Seeed Development Limited*

Writing Your First Arduino Sketch for PLCs

Arduino software can be used to write software for PLCs with its core framework without using a specific PLC-oriented Arduino software library.

Listing 4-1 shows the code for an Arduino sketch that can be used to control the **Grove LED** with a **Grove button**.

Listing 4-1. PLC with Push Button and LED Example (plc_1.ino)

```
int GROVE_BUTTON = 2; //Grove Button connected to D2
int GROVE_LED = 3; //Grove LED connected to D3

void setup(){
pinMode(GROVE_BUTTON, INPUT); //set button as an INPUT device
pinMode(GROVE_LED, OUTPUT); //set LED as an OUTPUT device
}
```

```
void loop(){
int buttonState = digitalRead(GROVE_BUTTON); //read the status of the button
if(buttonState == 1) //if button is pressed
digitalWrite(GROVE_LED,1); //turn on the LED
else
digitalWrite(GROVE_LED,0); //if button is released
}
```

Uploading Your Arduino Sketch

Now you're ready to upload your first PLC sketch to the Arduino board. Before that, connect your hardware setup with the computer using a **USB type A/B cable**. Use the following steps to upload the Arduino sketch:

1. Select the board type as **Arduino/Genuino**.

2. Select the correct **COM port.**

3. Click the **Verify** button to compile the sketch.

4. Click the **Upload** button to upload the sketch to the Arduino board.

Testing Your Sketch

Before testing anything, it's a best practice to supply power from an *external power source* because it can provide more current (*amperes*) than the USB port.

Table 4-5 shows some test cases that can be used to test your PLC.

***Table 4-5.** Test Cases*

Grove Button	Grove LED
IDEAL	OFF
PRESSED	ON
RELEASED	OFF

The **Grove Button** is equivalent to a *momentary switch* or a *push button,* which only remain in their **ON** state as long as they're being pressed. Note that the ideal state of the *Grove button* is **OFF**.

You can use the following steps to test your first Arduino-based PLC hardware setup with the embedded software.

1. In the ideal state of the *Grove button,* the *Grove LED* should be **OFF**.

2. Press the *Grove button* and *hold.*

3. Release the *Grove button* to see whether the *Grove LED* gets turned off.

Troubleshooting

A best practice is to remove any power sources attached to the Arduino before starting to troubleshoot. Here are some troubleshooting tips that you can use to troubleshoot your hardware setup.

1. Check the connections between the Arduino and the Grove Shield. Is the Grove Shield properly seated on the Arduino? Try to find any misaligned or bent wire-wrap headers.

2. Check the connections between the Grove connectors. Are they properly connected?

3. Verify the Grove port numbers mentioned in the Arduino sketch. Correct them according to the physical connections.

4. Check the *VCC power switch*. Slide it to the 5V position.

Working with Audio

Let's add another output device that is a small speaker to the PLC. It can work in parallel to the LED output. A **speaker** is suitable for hearing an output, or it could be used in parallel to any output.

Connecting the Components

Connect a **Grove speaker** to the *connector* marked **D4** (*Grove port number 4*) using a **Grove cable**. The *connector* **D4** internally connects to *Arduino digital pin 4* (Figure 4-11).

Figure 4-11. *Grove speaker connected to a Grove port D4. Image courtesy of Seeed Development Limited*

Now you can modify the first Arduino sketch listed in Listing 4-1 to interface with the new output device, which is the Grove speaker. The speaker should produce a short **BEEP** for a limited time when you press the **push button**.

The additional statements you need for this sketch are listed in bold in Listing 4-2.

Listing 4-2. PLC with Push Button, LED, and Speaker Example (plc_2.ino)

```
int GROVE_BUTTON = 2; //Grove Button is connected to D2
int GROVE_LED = 3; // Grove LED is connected to D3
int GROVE_SPEAKER = 4; // Grove Speaker is connected to D4

void setup(){
pinMode(GROVE_BUTTON, INPUT); //set button as an INPUT device
pinMode(GROVE_LED, OUTPUT); //set LED as an OUTPUT device
pinMode(GROVE_SPEAKER, OUTPUT); //set Speaker as an OUTPUT device

}

void beep(){
digitalWrite(GROVE_SPEAKER,1);
delay(10);
digitalWrite(GROVE_SPEAKER,0);
}

void loop(){
int buttonState = digitalRead(GROVE_BUTTON); //read the status of the button
if(buttonState == 1){ //if button is pressed
digitalWrite(GROVE_LED,1); //turn on the LED
beep(); //make audible tone, a beep
}
else{
digitalWrite(GROVE_LED,0); //if button is released
}
}
```

Upload the new sketch to the Arduino board to overwrite the previous sketch. Make sure to select the correct *board type* and *COM port* before uploading. Once it's uploaded to the Arduino, connect it to an external power supply and start testing.

Testing Audio

Press the **momentary push button** and hold. You will hear a short **BEEP** from the **speaker** and the **LED** will **light up** until you release the button. The **on-board potentiometer** can be used to adjust the loudness of the speaker.

You can improve this sketch by alternating the sound produced by the Arduino using the library. Visit https://code.google.com/archive/p/rogue-code/wikis/ ToneLibraryDocumentation.wiki to get more information about the Arduino tone library and to download the latest version.

Adding a Reset Button

Adding a *medium-sized* external **RESET** button to your PLC will help you quickly and easily restart any automated process (Figure 4-12). If you are planning to implement this PLC with an enclosure, an external reset button can be used to access the reset function.

To add an extra reset button in parallel with the on-board reset button, you need the following hardware components.

- Momentary push button (https://www.adafruit.com/products/1477)

- Two wires (red and black)

Connecting the Components

Follow these instructions to add the button to the **Grove Base Shield**:

1. Solder two wires to the *reset button,* as marked in Figure 4-12.

Figure 4-12. *Soldering the two wires for the reset button. Image courtesy of Adafruit Industries*

2. Then connect the red wire to the Arduino **RESET** pin.

3. Finally, connect the black wire to the Arduino **GND** pin.

Testing the Reset Button

Power up your Arduino PLC with an external power source, then press and release the newly connected reset button. The *orange color LED* connected to pin 13 should flash on and off. This will indicate that the reset button is working correctly.

Summary

In this chapter, you learned and grasped some hands-on experience in basic Arduino-based PLC by working with digital inputs and outputs. In the next chapter, you will learn how to build a simple PLC with **ArduiBox** and work with analog inputs.

CHAPTER 5

■ ■ ■

Building with an ArduiBox

In the previous chapter, you learned about how to build a basic Arduino-based PLC with the help of a **Grove Base Shield**. If you are planning to install your Arduino-based PLCs in an industrial environment, you should protect them from physical damages that can frequently occur in industrial environments. Also, the connections of the PLC should be easily accessible to quickly connect industrial peripherals, and to connect to external power supplies with high amperage.

ArduiBox

The ArduiBox provides an *industrial grade* set of accessories and components to build a PLC that can be used in industrial environments. The software can be written with Arduino software or any PLC libraries that support Arduino software. Here is a list of the parts you get with the ArduiBox package.

- Milled cab rail enclosure

- Transparent top shell

- Prototyping plate (main board)

- 4x 2-pin terminal blocks

- 2x 3-pin terminal blocks

- Sockets for the Arduino UNO, 101 and zero (male headers)

- Sockets for an optional shield (female headers)

- 2x self-tapping screws

- Reset button (optional)

- Components for power supply and voltage regulator (optional)

- 3x DIN rail mounting clips

Figure 5-1 shows the accessories and components you get to construct the **ArduiBox, Basic Version**. A DIN rail is not included with the ArduiBox package, so you should purchase one with mounting accessories. They are available at hardware stores or you can purchase one online.

© Pradeeka Seneviratne 2017

P. Seneviratne, *Building Arduino PLCs*, DOI 10.1007/978-1-4842-2632-2_5

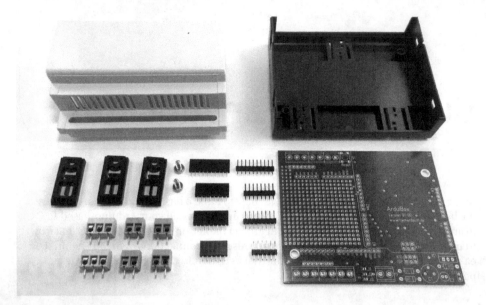

Figure 5-1. *Accessories/components of the ArduiBox. Image courtesy of Hartmut Wendt* (www.hwhardsoft.de)

The **ArduiBox, Standard Version** does include components (Figure 5-2) for building the on-board *power supply* and *voltage regulator* circuit.

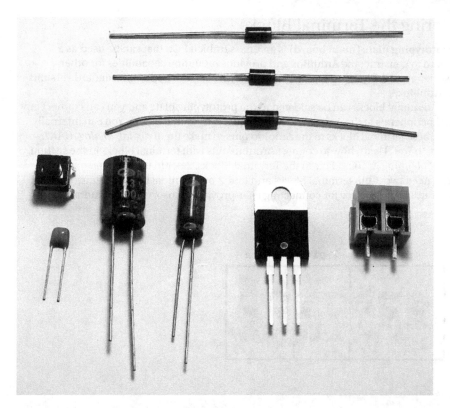

Figure 5-2. *Components for the power supply and voltage regulator. Image courtesy of Hartmut Wendt (www.hwhardsoft.de)*

Now let's begin to construct an ArduiBox-based PLC step-by-step.

Soldering the Terminal Blocks

The **prototyping plate** (**main board**) is an unassembled PCB that can be used as a baseboard to construct the ArduiBox and provides mounting capabilities for other accessories as well. The prototyping plate is identical on the basic and standard versions of the ArduiBox.

The terminal blocks can be soldered to the prototyping plate and you can connect any external peripherals to the Arduino UNO through the terminal blocks. You can internally connect these terminal blocks to the Arduino pins to make them *digital, analog, UART, or I2C ports*. You'll learn how to connect Arduino pins with terminal blocks in the section entitled, "Mapping Arduino Pins to the Terminal Blocks," later in this chapter.

You need two 3-pin terminal blocks and four 2 pin terminal **blocks** (Figure 5-3) to build the external interface for connecting user-provided sensors and actuators.

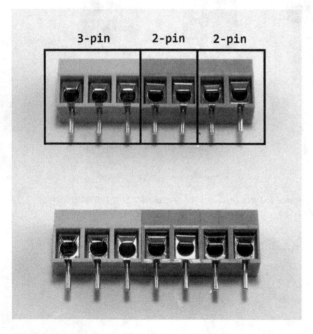

Figure 5-3. *3-pin and 2-pin terminal blocks. Image courtesy of Hartmut Wendt* (*www.hwhardsoft.de*)

Figure 5-4 shows two soldering areas for terminal blocks. Each row requires one **3-pin terminal block** and two **2-pin terminal blocks**. The mounting area for each terminal block is marked with a rectangle on the PCB. Now insert the terminal blocks into the PCB and solder them using a *soldering iron*. Make sure to place the wire face to the outside and solder them without creating any *cold soldering joints*.

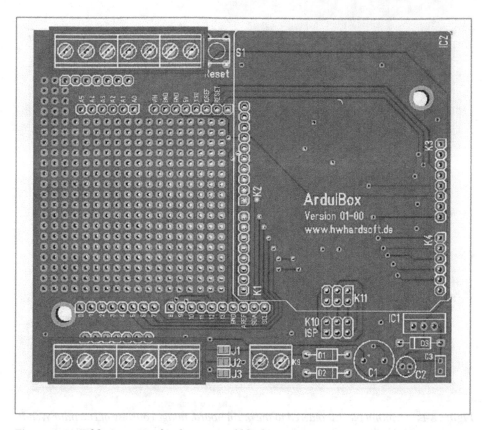

Figure 5-4. *Soldering areas for the terminal blocks*

Figure 5-5 shows the main board with the terminal blocks soldered.

Figure 5-5. *Terminal blocks soldered to the main board. Image courtesy of Hartmut Wendt* (www.hwhardsoft.de)

Soldering the Male Headers

To complete this task, you need four **male headers** (Figure 5-6) with a different number of pins.

Figure 5-6. *Male headers with 6, 8, and 10 pins*

Here is a list of male headers with a different number of pins required to build the Arduino UNO mounting area.

- 1 x 6-pole male header

- 2 x 8-pole male headers

- 1 x 10-pole male header

Now solder the male headers to the main board, as shown in Figure 5-7.

Figure 5-7. *Soldering the male headers. Image courtesy of Hartmut Wendt (www.hwhardsoft.de)*

These male headers can be used to mount the Arduino UNO board on the ArduiBox main board. Carefully mount the Arduino UNO board onto the ArduiBox main board and make sure the Arduino is properly seated on the male headers (Figure 5-8).

Figure 5-8. *Arduino UNO mounted on the ArduiBox main board. Image courtesy of Hartmut Wendt (www.hwhardsoft.de)*

Soldering the Female Headers

If you are planning to use a **shield** with the Arduino UNO board, you should solder the female headers to the *shield mounting area* of the ArduiBox main board. A shield can be used with Arduino to extend some functionalities, such as enabling Ethernet, driving relays, and many more.

ArduiBox comes with the following headers with a different number of pins.

- 1 x 6-pole female header

- 2 x 8-pole female headers

- 1 x 10-pole female header

Solder them to the main board marked with the *shield mounting area* **K1, K2, K3, and K4** (Figure 5-9).

- K1: 8-pole female headers
- K2: 10-pole female header
- K3: 8-pole female headers
- K4: 6-pole female headers

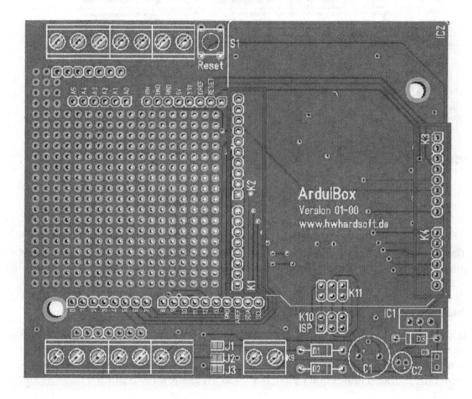

Figure 5-9. *Solder pads for Arduino shield mounting (shield mounting area)*

After soldering the female headers, the main board should be similar to the image shown in Figure 5-10.

93

Figure 5-10. *Female headers soldered to the main board. Image courtesy of Hartmut Wendt (www.hwhardsoft.de)*

Now plug the **Grove Base Shield** (or *any Arduino UNO compatible shield,* such as *Tinkerkit Shield, Ethernet shield,* or *Relay shield* per your requirements) into the **ArduiBox** *main board* using the *soldered female headers.*

Soldering the Reset Button

It is difficult to access the **Arduino UNO** *reset button,* because it is mounted upside down on the ArduiBox main board. As a solution, the ArduiBox main board provides a location to assemble an extra **reset button**, which is easy to access through the enclosure. Now solder the provided **momentary push button** to **S1,** as shown in Figure 5-11.

Figure 5-11. *A reset button assembled to the main board. Image courtesy of Hartmut Wendt (www.hwhardsoft.de)*

Now place and solder the **diode IN5819,** as shown in Figure 5-12.

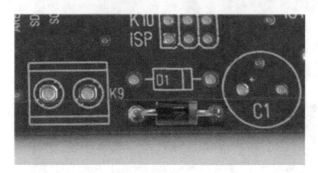

Figure 5-12. *IN5819 diode soldered to the main board. Image courtesy of Hartmut Wendt (www.hwhardsoft.de)*

Mapping Arduino Pins to the Terminal Blocks

To connect any input or output device to the terminal blocks, first you should internally make connections between the *Arduino pins* and the terminal blocks. Generally, input and output devices have two or more wires that need to be interfaced with Arduino.

Let's see how you can configure the terminals to connect a **temperature sensor** (*analog input*) and a **fan** (*digital output*).

The **temperature sensor** has three wires—**Power, Ground**, and **DATA**. Also, the **fan** has two wires—**Power** and **Ground**. Temperature sensors are typically connected to an Arduino analog input. Fans can be connected to a digital output. The fan should be turned on if the temperature is *equal to* or *greater than* 50 Celsius. Therefore, the Arduino has *two output* states for the fan, **0V** or **5V**.

As you can see in Figure 5-13, there are *eight solder pads* placed next to each row of the terminal blocks. Every terminal is internally connected with a soldering pad, respectively.

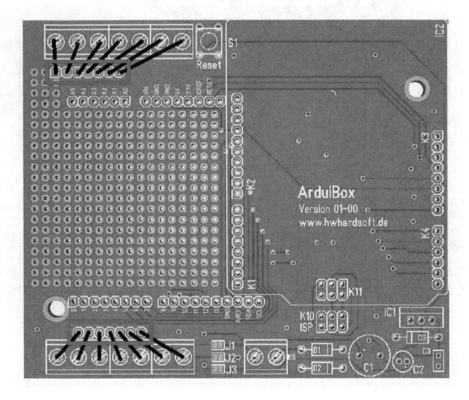

Figure 5-13. *PCB track connection between terminal blocks and solder pads*

Now make few connections between *Arduino pins* and *solder pads* using the **hook-up wires,** as shown in Figure 5-14 for the suite with our two input and output devices. The easiest way is to solder two 8-pole female pin headers to each soldering pad row before connecting them to the hook-up wires. If not, you can solder them permanently to the solder pads, but the main disadvantage is that you have to de-solder them for use with different Arduino pins.

The *temperature sensor* is connected to the *Arduino analog pin* **A0, 5V**, and **GND**. Also, the *fan* is connected to the *Arduino digital pin* **13** and **GND**.

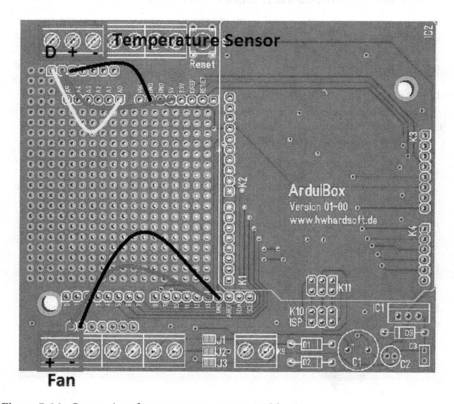

Figure 5-14. *Connections for temperature sensor and fan*

You'll need additional circuits to interface the temperature sensor and fan with the Arduino, depending on their technical characteristics.

97

Prototyping Area

The **prototyping area** of the *main board* is a very good place to build your own small circuits. You can use it to permanently solder your own circuits or build prototyping circuits by sticking a **self-adhesive breadbaord** (https://www.sparkfun.com/products/12043) on top of the prototyping area. Here are some technical specifications for the breadboard.

- Dimensions: (L x W x H) 46 x 35 x 9mm

- Contacts: 170 contacts in 17 columns and 10 rows

- Voltage ratings: <25V AC or < 60V DC circuits only

Building the Circuit

The **fan** we are going to use with the PLC requires about **180mA** of current, but the Arduino digital pin can provide a maximum of **40mA**. A driver circuit can be used to drive high loads from an Arduino digital pin with an additional power supply.

To build the circuit, you need the following components.

- TMP36 temperature sensor

- 5V DC brushless fan (https://www.jameco.com/z/KDE0503PEB1-8-SUNON-5-VDC-30mm-Brushless-Tubeaxial-Fan_2208905.html)

- TIP120 transistor (for the fan driver circuit)

- IN4004 diode

- 5V power supply of about 200mA or more

Figure 5-15 shows the pin layout of the **TMP36 temperature sensor**.

Figure 5-15. *TMP36 pin layout*

Figure 5-16 presents the circuit diagram for the **TIP120**-based *fan driver*. The datasheet of **TIP120** can be found at `https://www.fairchildsemi.com/datasheets/TI/TIP120.pdf` for your reference.

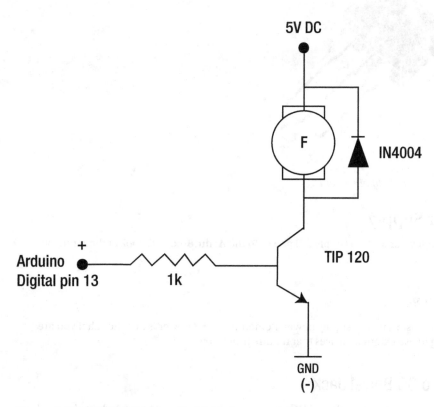

Figure 5-16. *DC motor driver circuit*

Figure 5-17 shows the pinout of the **TIP120** transistor.

Figure 5-17. *TIP120 pin layout*

Power Supply

There are several ways to supply DC power to the **ArduiBox**. Let's look at them one by one.

USB Power

This is the easiest way to supply power. Remember, USB power is not enough if you are planning to use external devices that require more power.

Arduino DC Barrel Jack

Supply the power using a **9V-12V DC** wall wart by connecting it to the **Arduino's DC barrel jack**.

Using a K9 Terminal

First solder a **2-pin terminal block** to a **K9** (Figure 5-18). With the K9 terminal block, you can use a **9-12V DC** power supply or a **15-30V DC** power supply to make the power. Also note that there are three jumpers—**J1, J2**, and **J3**—placed *adjacent* to the **K9 terminal block** that can be used to make necessary power paths per your requirements.

Figure 5-18. *K9 terminal block and three jumpers—J1, J2, and J3. Images courtesy of Hartmut Wendt (www.hwhardsoft.de)*

If you are planning to use a **9-12V DC** power supply with your **ArduiBox**, solder the **jumper J3** to create a *bridge*. In this case, you do not need to assemble the voltage regulator circuit on the main board.

However, if you want to use a **15-30V DC** power supply, you must first *assemble* the voltage regulator circuit on the main board (Figure 5-19). You do not need to solder the jumpers and can keep them all open, as is.

Figure 5-19. *Voltage regulator assembly. Images courtesy of Hartmut Wendt (www.hwhardsoft.de)*

Assembling the Enclosure

The ArduiBox enclosure consists of two parts:

- Bottom shell
- Top shell

You need **two** *self-tapping screws* to mount the main board to the **bottom shell**. Use a cross-slot (Phillips head) screwdriver to fasten the screws (Figure 5-20).

Figure 5-20. *Assembling the bottom shell. Image courtesy of Hartmut Wendt* (*www.hwhardsoft.de*)

On the backside of the **bottom shell**, there are three sockets for **DIN rail holders**. Mount them carefully from the inner channel to the outside (Figure 5-21).

Figure 5-21. *DIN rail clips attached to the bottom shell. Image courtesy of Hartmut Wendt* (*www.hwhardsoft.de*)

Finally, mount the **top shell** to complete the enclosure (Figures 5-22 and 5-23).

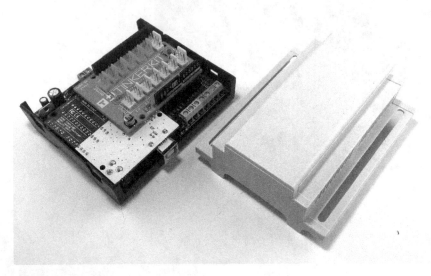

Figure 5-22. *The completed ArduiBox. Images courtesy of Hartmut Wendt* (www.hwhardsoft.de)

Figure 5-23. *The completed ArduiBox with the top shell assembled. Images courtesy of Hartmut Wendt* (www.hwhardsoft.de)

If you are planning to use the **Grove Base Shield** with an **ArduiBox**, carefully remove the top lid of the top shell using a small, flat screwdriver. This way, you can connect Grove devices to the Grove Base Shield by accessing things through the top shell.

DIN Rails

DIN rails are heavily used in PLC-based process automation systems, such as passive hardware, and allow you to mount PLCs and other related modules.

DIN rail and **DIN rail mounting track** are two common terms used in the industry for this same accessory; they can be used interchangeably. A DIN rail typically consists of the following accessories:

- Mounting track

- Mounting bracket plate

- End plate

- Spacers

- Hold down clip

Securely connect the ArduiBox to the DIN rail using DIN rail holders.

Connecting the Temperature Sensor and Fan

Now it's time to connect the temperature sensor and fan to your newly built ArduiBox PLC. Refer to Figure 5-14 before connecting them.

1. Connect the **TMP36** temperature sensor to the terminal blocks marked with the text **Temperature Sensor**.

   - Connect pin 1 (+Vs) of the temperature sensor to the ArduiBox terminal marked with +.

   - Connect pin 2 (Vout) of the temperature sensor to the ArduiBox terminal marked with D.

   - Connect pin 3 (GND) of the temperature sensor to the ArduiBox terminal marked with -.

2. Connect the **TIP120** fan driver to the terminal blocks marked with the text **Fan**.

   - Connect the positive lead (+) of the fan driver to the ArduiBox terminal marked with +.

   - Connect the negative lead (+) of the fan driver to the ArduiBox terminal marked with -/GND.

3. Connect a separate **5V-regulated power supply** to the fan driver between **5V DC** and **GND**.

4. Connect the ArduiBox to the computer's USB port using a USB type A/B cable.

5. Now verify and upload the Arduino sketch shown in Listing 5-1 to the Arduino board (now the ArduiBox). Select the correct board type and COM port as usual before uploading.

Listing 5-1. Temperature Reading Example (Temperature.ino)

```
int TEMPERATURE_SENSOR = A0;
int FAN   = 13;

void setup()
{
  pinMode(TEMPERATURE_SENSOR, INPUT);
  pinMode(FAN, OUTPUT);
}

void loop()
{
 int reading = analogRead(TEMPERATURE_SENSOR);

 float temperature = ((reading * 5.0) - 0.5) * 100 ;

  if (temperature >= 50)
  digitalWrite(FAN,HIGH);
  else
  digitalWrite(FAN,LOW);

 delay(1000);
}
```

6. Finally, remove the USB cable from the ArduiBox. Power it using an external power supply, as discussed in the section entitled "Power Supply".

Testing Your ArduiBox

You can use the following two test cases to test the Arduino sketch-defined temperature threshold level with the TMP36 temperature sensor.

Test Case 1

Apply heat to the TMP36 temperature sensor using a hair dryer or similar equipment. Depending on the threshold temperature level of *50 Celsius or greater than 50 Celsius*, the Arduino digital **pin 13** outputs a **HIGH** signal to the fan driver and the fan should turn on.

Test Case 2

Now apply cold to the TMP36 using ice cubes. When the temperature becomes *less than 50 Celsius*, the Arduino digital **pin 13** outputs a **LOW** signal and the fan should turn off.

Summary

In this chapter, you learned how to build a PLC by using Arduino UNO, Arduino shields, and an ArduiBox, which can be used in an industrial environment. In the next chapter, you will learn how to build and write PLC-style applications with ladder logic diagrams and plcLib—an Arduino software library.

■ ■ ■

CHAPTER 6

Writing PLC-Style Applications with plcLib

In the previous chapters, you built simple software applications for PLCs using Arduino software and its built-in libraries. When you're using Arduino built-in libraries to write applications for PLCs, you may encounter the following difficulties and disadvantages.

- A large amount of code is needed to perform a function

- Sketches become too complex with nested logic

- Debugging and fixing code is difficult

plcLib simplifies these difficulties by providing easy-to-use functions that can be used to write programs for PLC with a minimum amount of code.

Introduction to the plcLib Library

plcLib is an Arduino software library that can be used to write control-oriented PLC software applications for Arduino boards. The library provides a host of functions to write applications for control devices in industrial environments.

Installing plcLib on Arduino

The following steps explain how to download and install the plcLib library on your Arduino IDE.

1. Download the latest plcLib library from http://www.electronics-micros.com/resources/arduino/plclib/plcLib.zip. By default, a file named plcLib.zip will appear in your computer's downloads folder. Check your computer's download settings before downloading any software from web.

© Pradeeka Seneviratne 2017

P. Seneviratne, *Building Arduino PLCs*, DOI 10.1007/978-1-4842-2632-2_6

2. Extract the downloaded ZIP file with your favorite compression software. (If you have WinRAR installed on your computer, right-click on the file and select Extract Here from the context menu. A folder named plcLib will be created in the same directory.

3. Copy the plcLib folder to the Arduino libraries folder.

4. Restart the Arduino IDE and verify that the new library is installed on Arduino successfully. You can do this by choosing File ➤ Examples. Check whether you can see the menu item plcLib.

The Default Hardware Configuration

The plcLib library provides software-defined inputs and outputs. At a minimum, there are four inputs and four outputs for Arduino UNO. If you're using an Arduino MEGA, you might have more I/O pins to work with plcLib.

Table 6-1 shows the mapping relationship between software-defined inputs and the Arduino UNO analog pins.

Table 6-1. *plcLib Input Mapping with Arduino*

plcLib Input	Arduino UNO Input Pin
X0	A0
X1	A1
X2	A2
X3	A3

The software-defined outputs are also mapped with Arduino digital pins, as shown in Table 6-2.

Table 6-2. *plcLib Output Mapping with Arduino*

plcLib Output	Arduino UNO Output Pin
Y0	3
Y1	5
Y2	6
Y3	9

Ladder Logic

Ladder Logic is the primary programming language of PLCs. It can be used to document the circuit logic and flow with a set of symbols, so anyone can easily understand.

Before PLCs were introduced, relay logic control systems were popular and used relay Ladder Logic diagrams to present and document the systems. Now PLC Ladder Logic is easier than relay logic and can be drawn with fewer symbols. It's easy to read and understand.

A simple switch circuit can be converted to a PLC Ladder Logic diagram through a series of evaluations. First, the simple switch circuit is converted to a relay circuit, then it's converted to the relay Ladder Logic circuit, and finally to a PLC Ladder Logic.

Basic Ladder Logic Symbols

The symbols shown in Figure 6-1 can be used to draw a basic **PLC Ladder Logic diagram** to represent the inputs and outputs of a system.

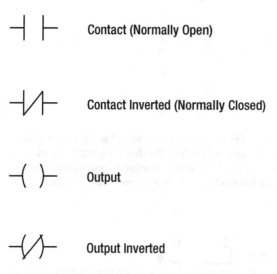

Contact (Normally Open)

Contact Inverted (Normally Closed)

Output

Output Inverted

Figure 6-1. Basic Ladder Logic symbols

Implementing Simple PLC-Style Applications

You can build PLC-style applications step-by-step by implementing switch circuits, relay circuits, relay Ladder Logic diagrams, and PLC Ladder Logic diagrams. Finally, the PLC Ladder Logic diagram can be transformed into an Arduino sketch by using plcLib library functions and other core Arduino functions. In this chapter, you learn about some basic and simple implementations. However, you can implement more complex applications with PLC Ladder Logic diagrams and the plcLib Arduino library.

111

Single Bit Input

Let's start with a simple switch circuit. The switch labeled S1 is controlling a lamp LS1, as shown in Figure 6-2.

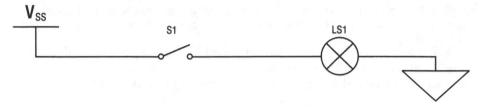

Figure 6-2. *Switch circuit*

The switch circuit action is described as, "The lamp LS1 is on when switch S1 is on (closed)". All possible combinations of the switch S1 and the lamp **LS1** action are shown as a truth table in Table 6-3.

Table 6-3. *Truth Table*

S1	LS1
Off	off
On	On

To implement this function using relays (Figure 6-3), the switch S1 is not connected to the lamp directly, but is connected to a relay coil labeled **RY1** that is normally open. The RY1 relay coil is used to control the relay coil RY2, whose contacts control the lamp. As you can see, the inputs and outputs are controlled by independent relay coils; in this case they are all normally open contacts.

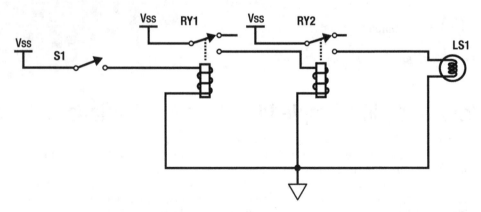

Figure 6-3. *Equivalent relay circuit*

The truth table can be rewritten with the actions of the two relay coils, as shown in Table 6-4.

Table 6-4. *Truth Table*

S1	RY1	RY2	LS1
Off	Off	Off	Off
On	On	On	On

The relay circuit can be converted further to a Relay Ladder Logic diagram, as shown in Figure 6-4.

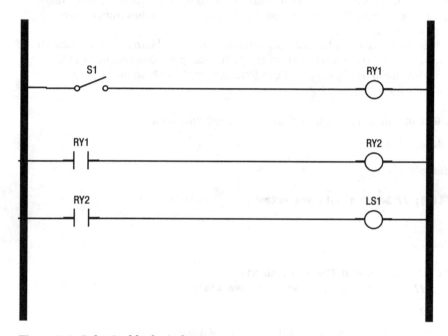

Figure 6-4. *Relay Ladder logic diagram*

The relay Ladder Logic diagram can be further converted to PLC Ladder Logic, as shown in Figure 6-5.

Figure 6-5. *PLC Ladder Logic*

A Ladder Logic diagram represents two vertical power **rails** (positive and negative) at each side, with horizontal circuit branches connected between the rails called **rungs**. A complex Ladder Logic diagram can have a series of circuit branches (rungs), which represents a separate circuit.

Further, the PLC Ladder Logic diagram can be transformed into an Arduino sketch using the plcLib library, as shown in Listing 6-1. This sample Arduino sketch can be opened with Arduino IDE by choose File ➤ Examples ➤ plcLib ➤ InputOutput ➤ BareMinimum from the menu bar.

Listing 6-1. Button Test with plcLib Example (BareMinimum.ino)

```
#include <plcLib.h>

void setup()
{
setupPLC(); // Setup inputs and outputs
}

void loop()
{
in(X0); // Read Input 0 (Read Switch S1)
out(Y0); // Send to Output 0 (Send to Lamp LS1)
}
```

Let's take a look at some important functions used in this sketch.

The #include <plcLib.h> imports the plcLib library and allows you to use its functions.

The setupPLC() function can be used to configure the PLC with the default input and output pin configuration for an Arduino UNO and Arduino MEGA.

The in() function reads the state of input X0 and the out() function sends output to Y0.

You can easily test this Arduino sketch with a Grove Base Shield. You'll need the following things to set up the hardware.

- Arduino UNO

- Grove Base Shield

- Grove Button

- Grove LED

Use the following steps to connect them, and then upload, execute, and test the code.

1. Connect the Grove Base Shield to the Arduino UNO.

2. Connect the Grove button to the Grove port **A0**.

3. Connect the Grove LED to the Grove port **D3**.

4. Now connect the Arduino UNO to the computer using a USB type A/B cable and upload the Arduino sketch onto the Arduino board.

5. After completing the upload, test the PLC by pressing and releasing the push button switch. Here are the test cases.

   - Button pressed ➤ LED on

   - Button released ➤ LED off

The plcLib-based code is very clear and easily maintainable. It's simpler than the equivalent code written with built-in Arduino functions, as shown in Listing 6-2.

Listing 6-2. Button Test Without plcLib Example (ButtonTestWOplcLib.ino)

```
const int buttonPin = A0;     // the number of the push button pin
const int ledPin = 3;         // the number of the LED pin

// variables will change:
int buttonState = 0;              // variable for reading the push button status

void setup() {
  // initialize the LED pin as an output:
  pinMode(ledPin, OUTPUT);
  // initialize the push button pin as an input:
  pinMode(buttonPin, INPUT);
}

void loop() {
  // read the state of the push button value:
  buttonState = digitalRead(buttonPin);

  // check if the push button is pressed.
  // if it is, the buttonState is HIGH:
  if (buttonState == HIGH) {
    // turn LED on:
    digitalWrite(ledPin, HIGH);
  } else {
```

```
    // turn LED off:
    digitalWrite(ledPin, LOW);
  }
}
```

In next section, you learn how to deal with inverted inputs and how to use them with PLC applications.

Inverted Single Bit Input

The same switch circuit we used in the section entitled "Single Bit Input" can be used to perform a different action, but the switch used with this circuit must have normally-closed contacts.

The switch labeled S1 is controlling a lamp LS1, as shown in Figure 6-6.

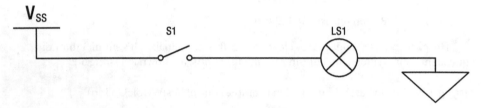

Figure 6-6. *Switch circuit*

The switch circuit action is described as, "The lamp LS1 is on when switch S1 is off (open)". All possible combinations of the switch S1 and the subsequent lamp LS1 action are shown as a truth table in Table 6-5.

Table 6-5. *Truth Table*

S1	LS1
Off	On
On	Off

Just as in the previous example, "Single Bit Input," this function can also be implemented using relays (Figure 6-7).

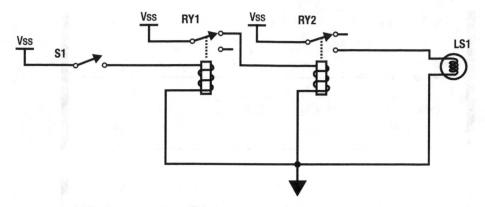

Figure 6-7. *Equivalent relay circuit*

The switch S1 is not connected to the lamp directly, but is connected to a relay coil labeled RY1 that is normally closed. The RY1 relay coil is used to control the relay coil RY2 that is normally open, whose contacts control the lamp. As you can see, every input and output is controlled by a relay coil; in this case, they are all normally open contacts.

The truth table can be rewritten with the two relay coils, as shown in Table 6-6.

Table 6-6. *Truth Table*

S1	RY1	RY2	LS1
Off	On	On	On
On	Off	Off	Off

The relay circuit further can be converted to a relay Ladder Logic diagram, as shown in Figure 6-8.

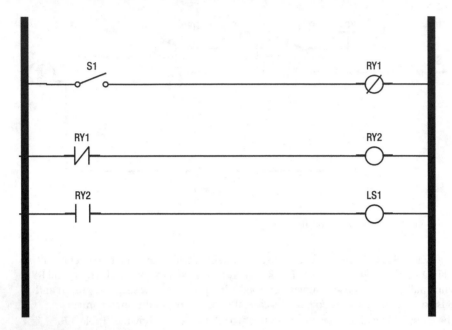

Figure 6-8. *Relay Ladder Logic diagram*

The relay Ladder Logic diagram can be further converted to PLC Ladder Logic, as shown in Figure 6-9.

Figure 6-9. *PLC Ladder Logic*

The equivalent Arduino sketch for the Ladder Logic diagram is shown in Listing 6-3. Note that the inNot() function can be used to invert the user input every time.

Listing 6-3. Inverted Single Bit Input Example (InvertedSingleBitInput.ino)

```
#include <plcLib.h>

void setup()
{
setupPLC(); // Setup inputs and outputs
}

void loop()
{
inNot(X0); // Read Input 0 (inverted) - read input Switch S1 (inverted)
out(Y0); // Send to Output 0 (Send to Lamp LS1)
}
```

Inverted Single Bit Output

Similarly, you can make inverted output with plcLib using the outNot() function.

The truth table for the switch, two relay coils, and the lamp is shown in Table 6-7.

Table 6-7. *Truth Table*

S1	RY1	RY2	LS1
On	On	Off	Off
Off	Off	On	On

The Arduino sketch shown in Listing 6-4 makes inverted outputs for every input it receives.

Listing 6-4. Inverted Single Bit Output Example (InvertedSingleBitOutput.ino)

```
#include <plcLib.h>

void setup()
{
setupPLC(); // Setup inputs and outputs
}

void loop()
{
in(X0); //read input Switch S1
outNot(Y0); // send to Lamp LS1 (inverted)
}
```

119

Time Delays

The plcLib library provides useful ways to implement time delays with industrial device control. The functions are as follows:

- timerOn()
- timerOff()
- timerPulse()

Turn On Delay

The timerOn() function delays activating an output until the input has been continuously active for the specified period of time in milliseconds.

The function accepts two parameters:

- Timer—Holds elapsed time
- Delay—Delay in milliseconds

The Arduino sketch (Listing 6-5) for this solution can be found at File ➤ Examples ➤ plcLib ➤ TimeDelays ➤ SwitchDebounce in your Arduino IDE.

Listing 6-5. Switch Debounce Example (SwitchDebounce.ino)

```
#include <plcLib.h>

unsigned long TIMER0 = 0;    // Define variable used to hold timer 0 elapsed time

void setup() {
  setupPLC();                // Setup inputs and outputs
}

void loop() {
  in(X0);                    // Read Input 0
  timerOn(TIMER0, 10);       // 10 ms delay
  out(Y0);                   // Output to Output 0
}
```

According to the sketch, the **X0** reads the input and provides output to **Y0** after 10ms.

When you press the push button, the LED will turn on after a 10ms delay. You can experiment with this again by increasing the delay.

```
timerOn(TIMER0, 3000);     // 3s delay
```

Turn Off Delay

Similar to the turn on delay, the timerOff() function delays the turning off of the output, when you turn off the input.

```
timerOff(timer, delay)
```

- Timer—Holds the request the elapsed time

- Delay—Delay in milliseconds (ms)

The plcLib library provides sample sketches to test the timerOff() function. Open the Arduino sketch (Listing 6-6) by choosing File ➤ Examples ➤ plcLib ➤ TimeDelays ➤ DelayOff.

Listing 6-6. Time Delay Example (DelayOff.ino)

```
#include <plcLib.h>

unsigned long TIMER0 = 0;   // Variable to hold elapsed time for Timer 0

void setup() {
  setupPLC();               // Setup inputs and outputs
}

void loop() {
  in(X0);                      // Read Input 0
  timerOff(TIMER0, 2000);   // 2 second turn-off delay
  out(Y0);                     // Output to Output 0
}
```

According to the sketch, the **X0** reads the input and provides output to **Y0**. From the circuit's point of view, when you press the push button, the LED will turn on after a 10ms delay. You can experiment with this again by increasing the delay.

```
timerOn(TIMER0, 3000);     // 3s delay
```

Fixed Duration Pulse Output

The timerPulse() function can be used to create a fixed-duration pulse output for a small duration input pulse.

Open the sample sketch (Listing 6-7) by choosing File ➤ Examples ➤ plcLib ➤ TimeDelays ➤ FixedPulse on the menu bar. The sample sketch code looks like Listing 6-7.

Listing 6-7. Fixed Duration Pulse Output Example (FixedPulse.ino)

```
#include <plcLib.h>

unsigned long TIMER0 = 0;     // Variable to hold elapsed time for Timer 0

void setup() {
  setupPLC();                   // Setup inputs and outputs
}
```

```
void loop() {
  in(X0);                      // Read Input 0
  timerPulse(TIMER0, 2000);    // 2 second pulse
  out(Y0);                     // Output to Output 0

}
```

The **X0** reads the input and provides a two-second pulse output to the **Y0**.

To test, press the Grove button and release it immediately. The LED will remain **ON** for 2 seconds and then turn **OFF**. If you can't do it with a two-second delay, just increase the delay to about five seconds.

Boolean Operations

PLC Ladder Logic can be used to implement Boolean logic functions such as AND, OR, XOR, NOT, NAND, NOR, and XNOR.

Among these Boolean logic functions, let's try to implement *Boolean OR operation* with PLC Ladder Logic and write a sketch using plcLib.

Implementing Boolean OR

Let's assume we have a circuit that consists of two normally open switches wired in parallel to control a lamp. The lamp is on when switch **S1** is on (closed) or when switch **S2** is on (closed). The truth table for this implementation is shown in Table 6-8.

Table 6-8. *Truth Table*

S1	S2	LS1
Off	Off	Off
Off	On	On
On	Off	On
On	On	On

Figure 6-10 shows the equivalent relay diagram.

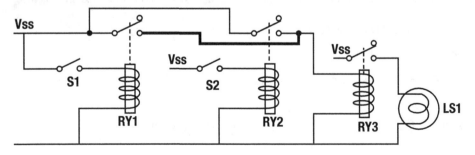

Figure 6-10. *Equivalent relay diagram*

The switch **S1** is connected to the normally open relay **RY1**, the switch **S2** is connected to the normally open relay **RY2**, and the lamp **LS1** is connected to the normally open relay **RY3**. The two relays are wired in a parallel series to the lamp.

The truth table for the switch, two relay coils, and the lamp is shown in Table 6-9.

Table 6-9. Truth Table

S1	S2	RY1	RY2	LS1
Off	Off	Off	Off	Off
Off	On	Off	On	On
On	Off	On	Off	On
On	On	On	On	On

The equivalent relay Ladder Logic diagram is shown in Figure 6-11.

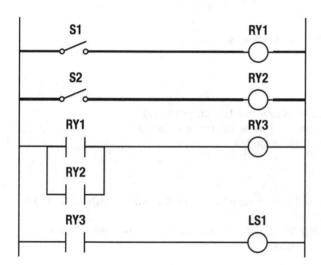

Figure 6-11. Equivalent relay Ladder Logic diagram

The diagram in Figure 6-11 can be further converted into the PLC Ladder Logic, as shown in Figure 6-12.

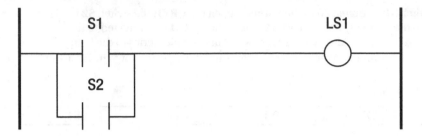

Figure 6-12. *PLC Ladder Logic*

Here is the Arduino sketch (Listing 6-8) for the Boolean OR logic function. First, you should add another Grove button to Grove port A1.

Listing 6-8. Boolean OR Logic Example (BooleanOrLogic.ino)

```
#include <plcLib.h>

void setup() {
  setupPLC();  // Setup inputs and outputs
}

void loop() {

  in(X0);      // Read Input 0 (Grove button on port A0)
  orBit(X1);   // OR with Input 1 (Grove button on port A1)
  out(Y0);     // Send result to Output 1

}
```

The function orBit(X1) is used to perform a *Boolean OR operation* with the input X0 and send the result to the output Y0.

Table 6-10 shows the complete list of plcLib functions that can be used to write Boolean operations with Arduino software.

Table 6-10. *plcLib Boolean Functions*

Boolean Operation	plcLib Function
AND	andBit()
OR	orBit()
XOR	xorBit()
NOT	outNot()
NAND	First andBit() then outNot()
NOR	First orBit(X1) then outNot(Y1)
XNOR	First xorBit(X1) then outNot()

You can perform AND, OR, XOR, and NOT Boolean operations using a single function, but the NAND, NOR, and XNOR Boolean operations require two functions to produce the correct output.

For an example, the NOR operation can be implemented as shown:

```
in(X0);      // Read Input 0
orBit(X1);   // OR with Input 1
outNot(Y1);  // Send result to Output 1 (inverted)
```

Summary

In this chapter, you learned about the Basic Ladder Logic symbols, relay Ladder Logic diagrams, and PLC Ladder Logic diagrams. You learned how to write Ladder Logic for simple applications with the plcLib Arduino library. In next chapter, you'll learn how to use the Modbus serial communication protocol with your PLC applications for transmitting information over serial lines between electronic devices.

CHAPTER 7

Modbus

Modbus is a *communication protocol* that can be used to *send* and *receive* **data** via a serial bus line, like **RS232** and **RS485** bus lines. In this chapter, you'll learn how to use Modbus communication protocol via an **RS485** bus line to connect industrial devices to your Arduino-based PLC. Modbus uses a **master-slave** *architecture,* where one *node* is configured as the **master** (i.e., Arduino PLC) and *other devices* are configured as **slaves** (temperature sensors, humidity sensors, light sensors, etc.). The advantage of using the RS485 is that it only uses *two shared wires* to connect all devices (slaves) to the master node. It also supports the use of devices in long distance and electrically noisy environments.

To connect your Arduino PLC to Modbus communication protocol enabled devices, first you should add some hardware modules to the Arduino in order to enable it as a Modbus master node.

The **RS485/Modbus module** is an ideal component that can be used to enable the Modbus communication protocol on your Arduino board. Additionally, you need a shield to connect and interface it with your Arduino. There are plenty of RS485/Modbus modules for Arduino available in the market.

To build the Arduino PLC enabled with RS485 and Modbus, you'll need the following things:

- Multiprotocol Radio Shield

- RS485/Modbus module

Multiprotocol Radio Shield

The **Multiprotocol Radio Shield** (Figure 7-1) from *Cooking Hacks* is an *Arduino UNO compatible shield* that's ideal for building Modbus-enabled PLCs. The shield is designed to connect to two communication modules at the same time.

© Pradeeka Seneviratne 2017
P. Seneviratne, *Building Arduino PLCs*, DOI 10.1007/978-1-4842-2632-2_7

Figure 7-1. *Multiprotocol Radio Shield from Cooking Hacks. Image courtesy of Libelium*
(https://www.cooking-hacks.com)

■ **Note** The multiprotocol shield has *two sockets* (Figure 7-2) that you can use to connect any hardware module that's **UART**-enabled. The two sockets are named SOCKET0 and SOCKET1. A socket consists of *2mm female pin headers,* which makes a total of *20 connections.* (**UART** stands for Universal Asynchronous Receiver Transmitter and it's one of the most popular serial protocols around.)

Figure 7-2. *Top view of a Multiprotocol Radio Shield. Image courtesy of Libelium*
(https://www.cooking-hacks.com)

All sockets are **SPI (Serial Peripheral Interface)** *enabled* so you can connect **RS458**, **RS232**, and *CAN Bus* **modules** to them. For **SOCKET0**, the SPI uses **3.3V** levels and for **SOCKET1**, the SPI uses **5V** levels.

There are **two** *wire-wrap headers* that are soldered on to the shield, so you can connect it with any Arduino UNO or compatible board. The shield physically connects to Arduino as follows.

- **Header 1**: Eight connections for Arduino digital pin, 0 to 7.

- **Header 2**: Six connections for Arduino analog pin, A0 to A5.

The shield also includes a **digital switch** to *enable* and *disable* the two sockets. You can control them using the software-defined library functions found in Arduino IDE.

RS485/Modbus Module for Arduino and Raspberry Pi

The **RS485/Modbus module for Arduino and Raspberry Pi** (Figure 7-3) allows you to connect more than one industrial devices to Arduino with only two wires. You can connect up to **32 devices** to Arduino using *two shared wires* by *addressing* each device with a unique identifier.

Figure 7-3. *RS485/Modbus module for Arduino and Raspberry Pi. Image courtesy of Libelium (https://www.cooking-hacks.com)*

Table 7-1 shows some technical specifications of the RS485/Modbus module for Arduino and Raspberry Pi, published by <u>Cooking Hacks</u>.

Table 7-1. *Technical Specifications of the RS485/ Modbus Module for Arduino and Raspberry Pi*

Standard	EIA RS485
Physical Media	Twisted pair
Network Topology	Point-to-point, multi-dropped, multi-point
Maximum Devices	32 drivers or receivers
Voltage Levels	-7V to +12V
Mark(1)	Positive voltages (B-A > +200mV)
Space(0)	Negative voltages (B-A < -200mV)
Available Signals	Tx+/Rx+, Tx-/Rx-(Half Duplex)Tx+,Tx-,Rx+,Rx-(Full Duplex)

Installing the RS485 Library for Arduino

To work with **RS485**, first you should install the **RS485 library for Arduino**. Use the following steps to install it on an Arduino IDE.

1. Extract the downloaded zip file, RS485_for_Arduino (see the "Modbus RS485 Library" section in Chapter 1 for more information about the downloading link). Use any compression software. You will get a folder called RS485_for_ Arduino.

2. The folder structure is very similar to the following hierarchy:

```
RS485_for_Arduino
            -> RS485
                            -> ModBusMaster485
                            -> ModbusSlave485
                            -> RS485
```

    Copy the ModBusMaster485, ModbusSlave485, and RS485 folders to your Arduino installation's libraries folder.

3. Finally, restart the Arduino IDE and verify whether you can see the sample sketches by choosing File ➤ Examples ➤ RS485.

If you can see them, you've successfully installed the RS485 library for Arduino.

Building a PLC with Modbus

Now you'll learn how to interface a temperature sensor with Arduino to use the Modbus communication protocol via an RS485 bus line. Then you will learn how to read sensor values from the temperature sensor and display them on an Arduino serial monitor.

Building the Hardware Setup

To build the hardware setup, you need the following things.

- Arduino UNO
- Multiprotocol Radio Shield
- RS485/Modbus module for Arduino
- TQS3-I MODBUS RS485 Interior Thermometer

The following steps go through the building process.

1. Connect the **Multiprotocol Radio Shield** to the Arduino UNO using wire-wrap headers (Figure 7-4**).**

2. Connect the RS485/Modbus module for Arduino and Raspberry Pi to **SOCKET 1** (Figure 7-4).

Figure 7-4. *An RS485 and Modbus-enabled Arduino setup. Image courtesy of Libelium (https://www.cooking-hacks.com)*

3. The **temperature sensor** we are going to use is **TQS3-I
    Modbus RS485 Interior Thermometer** (Figure 7-5) from
    PAPOUCH (www.papouch.com). It supports **Modbus** and
    Spinel *communication protocols* via an **RS485 bus line**.
    You can download the documentation for this product from
    http://www.papouch.com/en/shop/product/tqs3-i-rs485-
    interior-thermometer/tqs3.pdf/_downloadFile.php.

Figure 7-5. *TQS3-I Modbus RS485 interior thermometer. Image courtesy of Papouch
(http://www.papouch.com)*

The **Wago terminal blocks** reside inside the enclosure and are used to connect the
power supply and the RS485. Figure 7-6 shows the Wago terminal blocks labeled with
power and RS485 connections.

-   To connect the power supply between 7-20V DC, use the + and -
    terminals.

-   To connect to the RS485 bus line, use the TX+ and TX- terminals.

Figure 7-6. Wago terminal blocks. Image courtesy of Papouch (http://www.papouch.com)

By default, this temperature sensor is configured to communicate with the **Spinel** protocol (http://www.papouch.com/en/website/mainmenu/spinel/). A simple jumper setup can be used to configure it for the **Modbus RTU protocol,** as shown in Figure 7-7, by *shorting* the **setup jumper**.

Figure 7-7. *Setup jumper is shorted to enable the Modbus RTU. Image courtesy of Papouch* *(http://www.papouch.com)*

Now connect the **temperature sensor** to the **RS485/Modbus module for Arduino and Raspberry Pi** with two wires, as shown in Figure 7-8.

1. Connect the TX+ terminal of the temperature sensor to the RS485 module's terminal marked with **A** (non-inverted signal).

2. Connect the TX- terminal of the temperature sensor to the RS485 module's terminal marked with **B** (inverted signal).

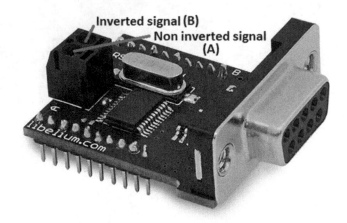

Figure 7-8. *Terminal for signal wires. Image courtesy of Libelium* *(https://www.cooking-hacks.com)*

3. Connect a DC power supply between 7-20V to the terminals marked with + and - using a wall wart power supply.

The Arduino Sketch

A ready-to-use Arduino sketch is available in the RS485 library. Use the following steps to modify the sketch according to your Modbus device.

1. Open your Arduino IDE and choose File ➤ Examples ➤ RS485 ➤ _RS485_04_modbus_read_input_registers to open the file named _RS485_04_modbus_read_input_registers.ino, as shown in Listing 7-1. The file will open in a new window.

Listing 7-1. Temperature Sensor Reading with Modbus Example (_RS485_04_modbus_ read_input_registers.ino)

```
#include <RS485.h>
#include <ModbusMaster485.h>
#include <SPI.h>

// Instantiate ModbusMaster object as slave ID 1
ModbusMaster485 node(254);

// Define one address for reading
#define address 101

// Define the number of bytes to read
#define bytesQty 2

void setup()
{

  // Power on the USB for viewing data in the serial monitor
  Serial.begin(115200);
  delay(100);
  // Initialize Modbus communication baud rate
  node.begin(19200);

  // Print hello message
  Serial.println("Modbus communication over RS-485");
  delay(100);
}

void loop()
{
  // This variable will store the result of the communication
  // result = 0 : no errors
  // result = 1 : error occurred
```

135

```
int result = node.readHoldingRegisters(address, bytesQty);

if (result != 0) {
  // If no response from the slave, print an error message
  Serial.println("Communication error");
  delay(1000);
}
else {

  // If all OK
  Serial.print("Read value : ");

  // Print the read data from the slave
  Serial.print(node.getResponseBuffer(0));
  delay(1000);
}

Serial.print("\n");
delay(2000);

// Clear the response buffer
node.clearResponseBuffer();

}
```

2. Now modify the value of the address variable in the Arduino sketch according to the address of your temperature sensor's holding register. See the product's datasheet to find the correct address for the holding register, as shown in Table **7-2**.

Table 7-2. Specifications of the Holding Register

Address	Access	Function	Description
102	Read	0x03	RAW value, which is the value as it was received from the sensors

In this example, the *temperature value* is stored in *address 102* and can be read with the function readHoldingRegisters().

// Define one address for reading
#define address 102

3. Modify the size of the register in bytes. In this example, the holding register at address 102 can store up to four bytes of data.

```
// Define the number of bytes to read
#define bytesQty 4
```

4. The following statement will store the result that's return by the readHoldingRegisters() function. The result of 0 indicates *no errors* and a result of 1 indicates that an *error has been occurred.*

```
int result = node.readHoldingRegisters(address, bytesQty);
```

5. To retrieve data from the response buffer, use following statement inside the loop() function.

```
getResponseBuffer(0)
```

6. You can print the data that is coming from the temperature sensor on an Arduino serial monitor using this statement:

```
Serial.print(node.getResponseBuffer(0));
```

7. After reading the data, don't forget to clear the response buffer. You can use the clearResponseBuffer(); function to clear the response buffer.

8. Now verify and upload the Arduino sketch to the Arduino board. After successfully uploading the sketch, open the Arduino serial monitor by choose Tools ➤ Serial Monitor. The serial monitor will print the values stored in the *response buffer,* as shown in Figure 7-9.

```
Modbus communication over RS-485
Read value: 213
Read value: 212
Read value: 214
Read value: 215
Read value: 213
Read value: 214
Read value: 213
Read value: 212
Read value: 214
Read value: 215
Read value: 213
Read value: 214
```

Figure 7-9. Arduino serial monitor output (readings of the response buffer)

The output values can be further converted to *Celsius* or *Fahrenheit* by using the conversion formulas mentioned in the datasheet of the temperature sensor. But some Modbus-enabled devices can directly output the required values in Celsius or Fahrenheit, without any further conversions.

Summary

In this chapter, you learned how to use the **Modbus communication protocol** via an *RS485 bus line* to connect industrial devices to your Arduino-based PLC and write Arduino sketches based on the *RS485 library for Arduino*. You also learned how to read values from devices that are enabled with the Modbus communications protocol. In the next chapter, you will learn how to map your Arduino-based PLCs into the *cloud* using a **NearBus Cloud Connector**.

CHAPTER 8

■ ■ ■

Mapping PLCs into the Cloud Using the NearBus Cloud Connector

In this chapter, you learn how to control your Arduino-based PLCs through the **Internet** by connecting them to a **cloud**. In simple terms, cloud computing is using a network of remote servers hosted on the Internet to store, manage, and process data, rather than a local server or a personal computer. A *cloud-connected PLC* can be controlled through the Internet with automated or manual inputs. You can also monitor your PLC through the Internet, such as the input data and output data, in real time, or you can use external services to control your PLC.

What Is NearBus?

Do you want to *synchronize* your Arduino board memory with cloud memory? Then this is the solution for memory mapping between Arduino and the cloud using the **NearBus** (www.nearbus.net) **cloud connector**. The memory mapping is done by mirroring or replicating a small part of Arduino's memory into the cloud's memory. So, reading or writing on the cloud's memory will have the same effect as reading or writing directly into the Arduino's memory. Also, the NearBus provides a set of web services known as **NearAPI** to control your Arduino board.

Building Your Cloud PLC

To build this project, you need the following things:

- Arduino UNO
- Arduino Ethernet shield
- Grove Base Shield
- Grove LED

© Pradeeka Seneviratne 2017
P. Seneviratne, *Building Arduino PLCs*, DOI 10.1007/978-1-4842-2632-2_8

- Ethernet cable

- USB type A/B cable

- 9V wall wart power supply

- Router with an Internet connection

- ArduiBox (optional; you can build this on ArduiBox)

Use the following steps to build the hardware setup:

1. Connect the **Arduino Ethernet shield** to the **Arduino UNO board** using the Ethernet shield's *wire-wrap headers*.

2. Connect the **Grove Base Shield** to the **Arduino Ethernet Shield** using the Grove Base Shield's *wire-wrap headers*.

3. Connect the **Grove LED** to **Grove port D3**.

4. Connect an **Ethernet cable** between the *Ethernet shield* and the *router*.

5. Connect an **Arduino UNO board** to the computer using the **USB type A/B cable**.

Mapping a PLC Into the Cloud Using NearBus Cloud Connector

Before connecting your PLC to the NearBus cloud, you should sign up with NearBus (http://nearbus.net/) and create a new account.

Signing Up with NearBus

Sign up is easy with **NearBus**; you simply provide a few basic details.

1. Click **Sign Up** on the NearBus *menu bar*.

2. Enter a valid *email address, username,* and *password*.

3. Click the **Sign Up** *button*.

A new user account will be created if you have provided valid information; otherwise correct the information and start from the beginning. After successfully creating a user account, NearBus will prompt you to *log in* with your credentials.

Defining a New Device in NearBus

After successfully logging into your NearBus account, you can define a new device on the cloud to map with your PLC.

1. Click **New Device** on the NearBus menu bar. The browser will load the **New Device Setup** page, as shown in Figure 8-1.

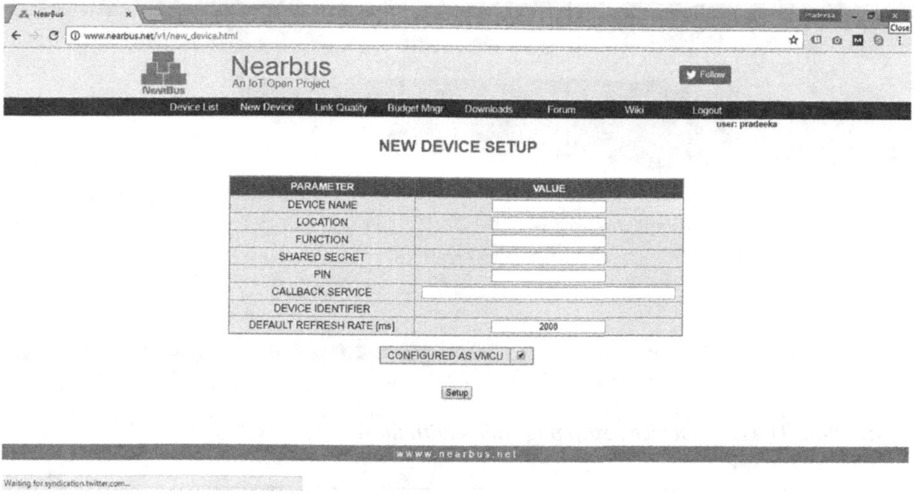

Figure 8-1. *The New Device Setup page*

2. On the **New Device Setup** page, provide suitable values to these parameters.

    - Device Name: *MY CLOUD PLC*

    - Location: *(blank)*

    - Function: *(blank)*

    - Shared Secret: *12345678*

    - PIN: *(blank)*

    - Callback Service: *(blank)*

    - Default Refresh Rate: *Just keep the default rate of 2000 (1000 should work)*

■ **Note** The shared secret is a mandatory field and the length of the field is eight characters. You can use any alphanumeric characters to build the shared secret string.

---

3. *Check* the **CONFIGURED AS VMCU** *checkbox* and click the **Setup** *button,* as shown in Figure 8-2.

141

Figure 8-2. *The New Device Setup page filled with data*

> ■ **Note** The **VMCU** mode (**Virtual Microcontroller Unit**) allows direct control of the basic **MCU** (**Microcontroller Unit**) features, such as **GPIO** (**General Purpose Input Output**) and **ADC** (**Analog to Digital Converter**) via a web services **API** (Application Programming Interface).

4. Now you can view your device under device list by clicking **Devices List** on the *menu bar,* as shown in Figure 8-3.

Figure 8-3. *View by device list*

Note that the NearBus assigned a **device ID** to your *new device*. Also, it indicates that the *physical device* (*your Arduino-based PLC*) is currently in the DOWN state. This is true because we still have not configured it and connected it to the Internet.

Downloading the NearBus Library for Arduino

Now we are ready to download and install the **NearBus library** to your Arduino UNO board.

1. Click **Downloads** on the *menu bar*. The **Downloads** *page* will appear with various software library options, as shown in Figure 8-4.

2. For **Arduino Ethernet Shield**, you need to choose **Arduino library for Ethernet - Alpha Release**.

Figure 8-4. *NearBus Downloads page*

3. Click **NearBusEther_v16.zip** to *download* the library in compressed format.

4. After downloading the file, *extract* it with your favorite compression software. You'll get a folder called NearBusEther_v16. Copy this folder to your Arduino installation's libraries folder.

5. Also download the **FlexTimer2** *library* from http://www.nearbus.net/downloads/FlexiTimer2.zip. Extract it and copy the extracted folder to the Arduino installation's libraries folder.

143

Uploading the Sketch

1. Now restart your Arduino IDE and choose File ➤ Examples
   ➤ NearBusEther_v16 ➤Hello_World_Ether on the menu bar.
   The Hello_World_Ether Arduino sketch listed in Listing 8-1
   will open in a new window.

Listing 8-1. LED Controlled with NearBus Cloud Example (Hello_World_Ether.ino)

```
//////////////////////////////////////////////////////////////////////////
// NEARBUS LIBRARY - www.nearbus.net
// Description: Hello World Example
// Platform:    Arduino Ethernet
// Status:      Alpha Release
// Author:      F. Depalma
// Support:     info@nearbus.net
//////////////////////////////////////////////////////////////////////////
// REVISION HISTORY
// v0.20 - 08-02-13 - Initial Release
// v0.21 - 25-04-13 - This release includes support for Arduino Mega 128 and
256 (ADC function)  - Contributor: Peter Huff.
// v0.3L - 02-05-13 - This release implements 32bits Registers  for TRNSP
mode (Reg_A & Reg_B) and support for Google Connector.
// v0.4L - 10-05-13 - This release includes support to: Ethernet, WiFi and
GPRS Arduino shields.
// v0.5L - 29-06-13 - This release includes support for IP Static Addressing
- Contributor: Craig.
// v0.6  - 02-08-13 - This release implement DNS Address Resolution and
PULSE_OUTPUT NearBIOS function.
// v0.7  - 20-08-13 - This release implements support for user defined
functions (MY_NBIOS_0).
// v0.9  - 17-09-13 - This release implements Enhanced Services (Alpha) -
TriggerInput() - DigitalCounter() - RmsInput() and ResetPort().
// v0.10 - 05-11-13 - This release implements support for Arduino YUN (it
requires IDE 1.5.4) and fixes some minors bugs.
// v0.11 - 12-11-13 - This release fixes a BUG in the MY_NBIOS service.
// v0.12 - 28-11-13 - The RMS_INPUT service is modified to deliver a mV
output value.
// v0.14 - 03-01-14 - This release Include supports for X-CONTROL.
// v0.15 - 11-01-14 - This release Include supports for X-CONTROL V2.
// v0.16 - 16-04-14 - This release fix a register debug bug.
//
//////////////////////////////////////////////////////////////////////////
// AGENT CONFIGURATION PARAMETERS  ( You only need to define these
parameters )
// IMPORTANT: this agent version is only supported on devices ID from
NB100246 onwards
//////////////////////////////////////////////////////////////////////////
```

```
char deviceId []     = "agent_id";          // Put here the device_ID generated
by the NearHub ( NB1xxxxx )
char sharedSecret[] = "agent_password";     // (IMPORTANT: mandatory
8 characters/numbers) - The same as you configured in the NearHu
byte mac[6]          = { 0x90, 0xA2, 0xDA, 0x0D, 0x21, 0xEA };
// Put here the Arduino's Ethernet MAC

// ADDITIONAL CONFIGURATION FOR STATIC IP ADDRESSING
#define  STATIC_IP 0                         // 1=>Static IP  0=>DHCP
byte ip[]            = { 192,168,1,10 };  // Your Arduino IP Address
byte subnet[]        = { 255,255,255,0 }; // Your Arduino IP Mask
byte gateway[]       = { 192,168,1,1 };   // Your Default Gateway (LAN Router)
byte gdns[]          = { 8,8,8,8 };        // Google DNS server


/////////////////////////////////////////////////////////////////////////////

/////////////////////////////////////////////////////////////////////////////
//  Includes
/////////////////////////////////////////////////////////////////////////////
#include <Ethernet.h>                        // Ether Specific Configuration
#include <NearbusEther_v16.h>                 // [REL]
#include <SPI.h>
#include <Servo.h>
#include <FlexiTimer2.h>

/////////////////////////////////////////////////////////////////////////////
//  Global Variables
/////////////////////////////////////////////////////////////////////////////
Nearbus Agent(0);

ULONG A_register[8];                          // Define the Tx Buffer (Reg_A)
ULONG B_register[8];                          // Define the Rx Buffer (Reg_B)
int retorno;

void AuxPortServices(void) {
    Agent.PortServices();
}


/////////////////////////////////////////////////////////////////////////////
//  BROWSER API REST COMMAND LINE (for JavaScript) - (use it for
troubleshooting)
//
//  http://nearbus.net/v1/api_vmcu_jsb/NB100***?user=****&pass=****&channel=
0&service=DIG_OUTPUT&value=1&method=POST&reqid=123456
//
//  user:          Your NearBus Web user
//  pass:          Your NearBus Web password
```

```
//   channel:        NearBus channel [0-3]
//   value:          Service value (if apply)
//   method:         GET (read) / POST (write)
//   reqid:          Transaction identifier (to match a request and its
response)
//
//   SUPPORTED API SERVICES
//   DIG_INPUT:      Digital Input - Input Range [0-1] - Method: GET
//   DIG_OUTPUT:     Digital Output - Output Range [0-1] - Method: POST/GET
//   ADC_INPUT:      ADC Analog Input - Output Range [0-1023] - Method: GET
//   PULSE _OUTPUT: Digital Output - Input Range [0-65535] in steps of 10ms
(max 655 seg) - Method: POST/GET
//   PWM_OUTPUT:     PWM Output calibrated for Servomotors - Input Range [800-
2200] - Method: POST/GET
//   DIG_COUNTER:    Pulse Counter / Accumulator
//   RMS_INPUT:      True RMS Meter
//   MY_NBIOS_0:     User defined function
//
//   For a detailed information please go to: http://goo.gl/Gxrcua
//
///////////////////////////////////////////////////////////////////////////

/*########################################################################
#########################################################################
###      MY_NBIOS CUSTOM FUNCTION CODE                         #######
#########################################################################
#####################################################################*/

///////////////////////////////////////////////////////////////////////////
// MOISTURE SENSOR - EXAMPLE OF CUSTOM CODE
///////////////////////////////////////////////////////////////////////////
// Product Web: http://www.seeedstudio.com/depot/grove-moisture-
sensor-p-955.html
// Wiki: http://seeedstudio.com/wiki/Grove_-_Moisture_Sensor
// Code Source: n/a
// Technical Spec: Humidity: 5% RH - 99% RH - Temperature -40°C - 80°C -
Accuracy: 2% RH / 0.5°C
// Support Shield: Base Shield V1.3 - Grove compatible - http://
seeedstudio.com/depot/base-shield-v13-p-1378.html
//
///////////////////////////////////////////////////////////////////////////
void Nearbus::MyNbios_0( byte portId, ULONG setValue, ULONG* pRetValue, byte
vmcuMethod, PRT_CNTRL_STRCT* pPortControlStruct )
{

    //**********************************
    // Reconfiguring Ports as I/O
    //**********************************
```

```
    if( pPortControlStruct->portMode != MYNBIOS_MODE )
    {
            PortModeConfig( portId, MYNBIOS_MODE );
    }


    //**********************************
    // Custom Function
    //**********************************
    // DEFAULT:      The default analog reference of 5 volts (on 5V Arduino
    boards) or 3.3 volts (on 3.3V Arduino boards)
    // INTERNAL:     An built-in reference, equal to 1.1 volts on the
    ATmega168 or ATmega328 and 2.56 volts on the ATmega8 (not available on
    the Arduino Mega)
    // INTERNAL1V1:  A built-in 1.1V reference (Arduino Mega only)
    // INTERNAL2V56: A built-in 2.56V reference (Arduino Mega only)
    // EXTERNAL:     The voltage applied to the AREF pin (0 to 5V only) is
    used as the reference.

    analogReference( DEFAULT );

    *pRetValue = (ULONG) analogRead( pPortControlStruct->anaPinId );

}

/*###############################################################################
################################################################################
###             END OF CUSTOM CODE         ####                            ###
################################################################################
################################################################################*/

/////////////////////////////////////////////////////////////////////////////
//    SETUP ROUTINE
/////////////////////////////////////////////////////////////////////////////
void setup(void)
{

    //*******************************
    // SERIAL INTERFACE INITIALIZATION
    //*******************************
    Serial.begin(9600);                             // Start serial library

    //*******************************
    // NEARBUS INITIALIZATION
    //*******************************
    Agent.NearInit( deviceId, sharedSecret );
```

```
//*******************************
// ETHERNET INITIALIZATION
//*******************************
if( STATIC_IP ){
    Ethernet.begin( mac, ip, gdns, gateway, subnet );
}
else {
    Ethernet.begin( mac );
}


//*******************************
// FLEXITIMER INITIALIZATION
//*******************************
    #if FLEXI_TIMER
        FlexiTimer2::set( INT_PERIOD, AuxPortServices );
// Call the port services routine every 10 ms
    FlexiTimer2::start();
    #endif

    delay(1000);
// Give the Ethernet shield a second to initialize

    pinMode(3, OUTPUT);
}

void loop()
{
int ret;

    Agent.NearChannel( A_register, B_register, &ret );

    if ( ret >= 50 )
    {
        Serial.println( "Rx Error" );
        // [50]  Frame Authentication Mismatch
        // [51]  Frame Out of sequence
        // [52]  Remote ACK Error
        // [53]  Unsupported Command
    }

/*

    /////////////////////////////////////
    // Example 1 - Analog Input
    // Mode: TRNSP
    /////////////////////////////////////
    A_register[0] = analogRead(0);                                   // PIN A0
*/
```

```
/*
    /////////////////////////////////
    // Example 2 - Digital Output
    // Mode: TRNSP
    /////////////////////////////////
    if( B_register[0] == 1 ){
        digitalWrite( 3, HIGH );                          // PIN D3
    }
    else {
        digitalWrite( 3, LOW );                           // PIN D3
    }
*/

/*
    /////////////////////////////////
    // Example 3 - PWM Analog Output
    // Mode: TRNSP
    /////////////////////////////////
    analogWrite( 3, B_register[0] );                      // PIN D3
    - This function DO NOT works with Servo Motors and require an input from
    0 to 255.
*/

}
```

2. Now *modify* the Arduino sketch shown in Listing 8-1 according to your PLC configuration with NearBus. Follow these steps to make the modifications:

   a) #define STATIC_IP: This value can be 0 or 1. Use 1 for a **static IP** and **0 for a dynamic IP** configuration. If you use 0, there is no need to provide the IP address for your Ethernet shield.

```
// ADDITIONAL CONFIGURATION FOR STATIC IP ADDRESSING
#define  STATIC_IP  0  // 1=>Static IP  0=>DHCP
```

   b) byte ip[]: This is your Arduino's static IP address and should be written as a comma-separated value. Also, it should be in the valid IP address range of your network.

```
byte ip[]          = { 192,168,1,10 }; // Your Arduino IP Address
```

   c) byte gateway[]: Your router's IP address.

```
byte gateway[]     = { 192,168,1,1 };  // Your Default Gateway (LAN Router)
```

3. Now verify the Arduino sketch. You will likely get the following error in the output window.

```
In file included from D:\APress\Author Templates\Author Templates\Chapter
8\Source Codes\Hello_World_Ether\Hello_World_Ether.ino:47:0:

C:\Users\Pasindu\Downloads\arduino-1.6.11-windows\arduino-1.6.11\libraries\
NearBusEther_v16/NearbusEther_v16.h:95:22: fatal error: WProgram.h: No such
file or directory

 #include <WProgram.h>

                      ^

compilation terminated.

exit status 1

Error compiling for boarerd Arduino/Genuino Uno.
```

You can solve this issue with a small fix to the NearbusEther_v16.h *header file* located in the NearbusEther_v16 *folder*.

Find the line #include <WProgram.h> and change it to #include <Arduino.h>. **Then** save the file. Now try to verify the sketch again. The sketch should now verify without any errors. You can upload it to the Arduino board by clicking on the Upload button.

4. Remove your Arduino PLC from the computer and connect it to the external power supply.

Controlling the Grove LED from the NearBus Cloud

1. *Click* **Devices List** on the **NearBus** *web page,* as shown in Figure 8-5.

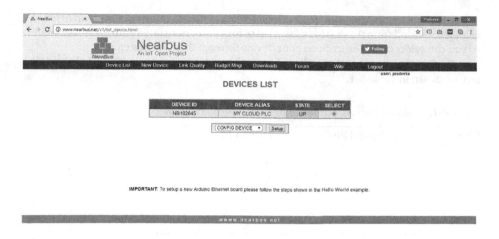

Figure 8-5. *The Devices List page*

The state column will show the current status of the Ethernet shield. The **UP** state indicates that your shield is successfully connecting and communicating with the NearBus cloud.

2. Select **Config Device** from the *drop-down* list (this is the default selection) and click the **Setup** *button*. The **Device Configuration** *page* will appear, as shown in Figure 8-6.

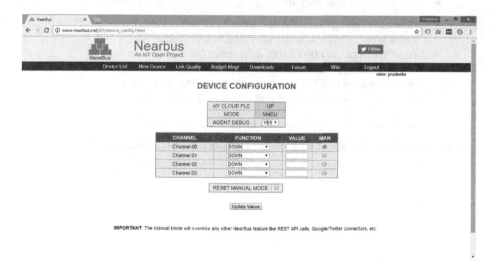

Figure 8-6. *The Device Configuration page*

151

The **Device Configuration** page enables you to control your device with *up to four channels,* labeled from 0 to 3. There are a set of functions associated with each channel, so you can select one to execute on your device through the NearBus cloud.

The NearBus has four channels and each channel is associated with an Arduino I/O pin, as shown in Table 8-1.

Table 8-1. *Arduino Pins Associated with Each NearBus Channel*

Channel	Digital Pin	Analog Pin
0	3	A0
1	5	A1
2	6	A2
3	9	A3

There is also a set of *functions* you can select form the **Function** *drop-down* list according to your requirements. Here are some of the functions you can use with your Arduino.

- UP: Digital output =1

- DOWN: Digital output =0

- DIG_INPUT: Digital input 0 or 1

- ADC_INPUT: 10 bits analog input (0 to 1023 for 0 to 1.1V)

- PWM_OUTPUT: PWM output

3. Now you're going to control the **Grove LED** connected to the *Grove port D3.* In channel 00, select the **MAN** *checkbox* and select **UP** from the **FUNCTION** *drop-down list,* as shown in Figure 8-7. The NearBus cloud will send the UP command to the Arduino and the Grove LED will turn on. This is the same as executing the function digitalWrite(3, HIGH) on digital pin 3.

■ **Note** By selecting the MAN checkbox, the channel will turn to manual mode.

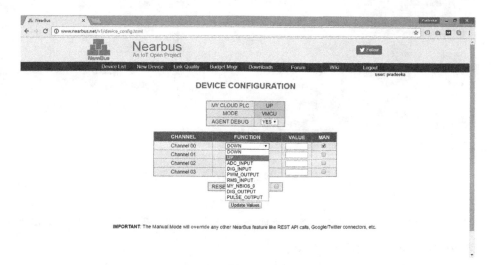

Figure 8-7. *The Device Configuration page*

4. Again, change the *function* to DOWN and see whether you can turn off the LED through the NearBus.

In this project, you'll learn how to control an Arduino I/O pin through the NearBus cloud using *manual inputs* (manual mode) like the UP and DOWN NearBus commands.

But in industrial environments, you should be able to control your Arduino-based PLC with various external inputs. Let's improve this project by adding an external clock to control the Grove LED by sending ON and OFF messages to the Arduino through the NearBus.

You can implement a clock using the **IFTTT** (www.ifttt.com) **DIY light platform**. The message strings for ON and OFF commands can be implemented using IFTTT and can be sent through a Twitter account to the NearBus cloud. Finally, the NearBus cloud will execute the commands on your Arduino.

Using the IFTTT DIY Light Platform

First you are going to configure your Arduino PLC with the Twitter Connector. The following steps explain how to do this:

1. In the **Devices List** page, as shown in Figure 8-5, select **Twitter Config** from the *drop-down* list and *click* the **Setup** button. The **Twitter Connector** page will appear, as shown in Figure 8-8.

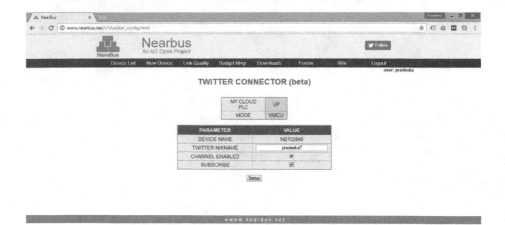

Figure 8-8. *The Twitter Connector page*

2. Type your **Twitter** account's name into the **Twitter Nickname** *text box*. Then check the **Channel Enabled** and **Subscribe** checkboxes. C*lick* the **Setup** button to continue.

3. Accept the request message sent by the NearBus in your Twitter account by *clicking* the **Accept** button.

Creating a Recipe with IFTTT

To proceed, you first need an **IFTTT user account**. If you don't have one, it's time to create an account by visiting https://ifttt.com.

1. On the IFTTT web page, shown in Figure 8-9, *click* **My Recipes** on the *menu bar*.

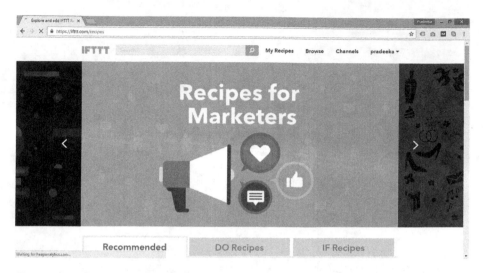

Figure 8-9. *The IFTTT recipes page*

2. Click the **Create a Recipe** *button,* as shown in Figure 8-10.

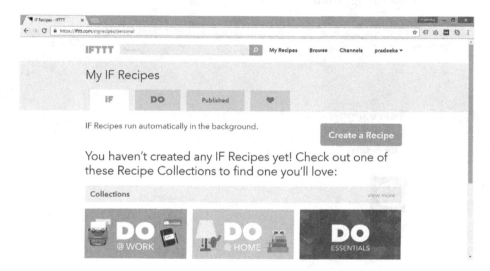

Figure 8-10. *Create a recipe*

3. *Click* the link shown in Figure 8-11.

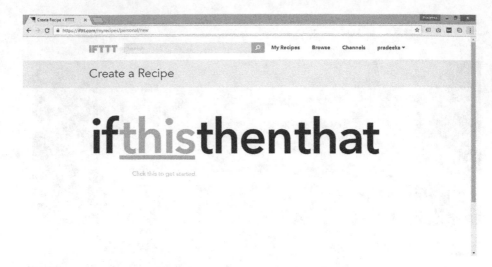

Figure 8-11. *Create a recipe: this*

4. *Search* for **date & time** and *click* the **Date & Time** icon, as shown in Figure 8-12.

Figure 8-12. *The Choose Trigger Channel page*

5. *Click* the **Every Hour At** option, as shown in Figure 8-13.

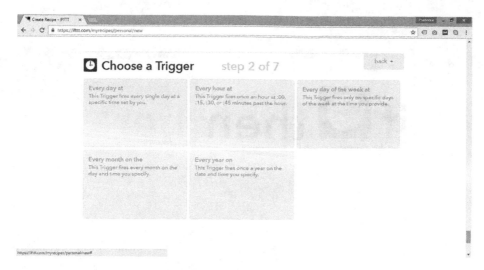

Figure 8-13. *The Choose a Trigger page*

6. From the **Minutes Past the hour** *drop-down list,* select the number of minutes past the hour to trigger a **Twitter message** (a **Tweet**). *Select* **00** *minutes* from the *drop-down list to fire once an hour* at **00** *minutes past the hour.* Then *click* the **Create Trigger** *button,* as shown in Figure 8-14.

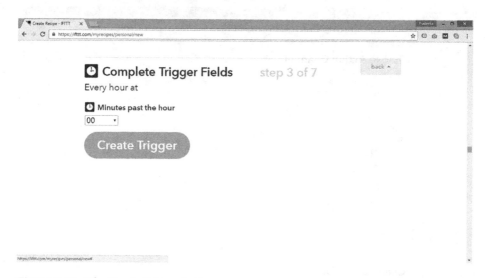

Figure 8-14. *The Create Trigger button*

7.   *Click the* that *link,* as shown in Figure 8-15.

Figure 8-15. *Create a Recipe: that*

8.   *Search* **Twitter** *for an* action channel. Then *click* the **Twitter** *icon,* as shown in Figure 8-16.

Figure 8-16. *Choose an action channel*

9. *Click* **the Connect** *button,* as shown in Figure 8-17.

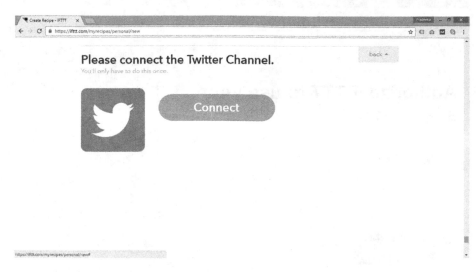

Figure 8-17. *Connecting to the Twitter channel*

10. *Click the* **Authorize App** *button* to authorize **IFTTT** to use your **Twitter** account to send tweets, as shown in Figure 8-18.

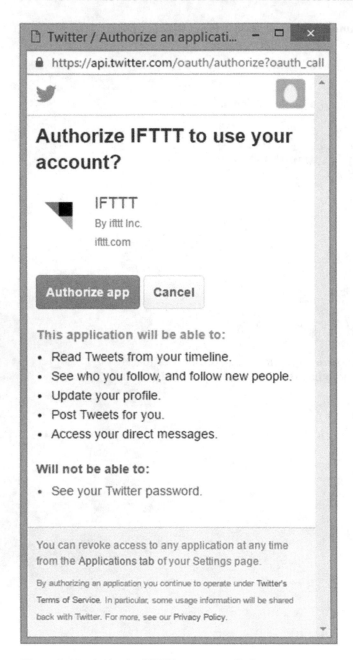

Figure 8-18. *Authorize IFTTT to use your Twitter account*

    11.   After successfully *authorizing* the **IFTTT**, you will get a successful message. *Click* **Done** to proceed.

12.   *Click* the **Continue to the Next Step** *button,* as shown in
       Figure 8-19.

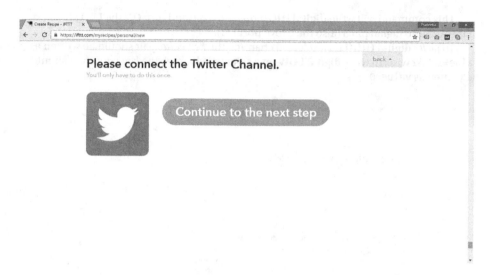

Figure 8-19. *A Twitter configuration step*

13.   *Click* **the Post a Tweet** *section,* as shown in Figure 8-20.

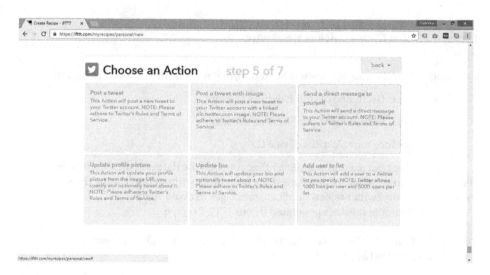

Figure 8-20. *The choose an action section*

14.   In the **Complete Action Fields** *page,* as shown in Figure 8-21,
       type the following command.

```
?user=****&pass=****&device=NB102xxx&channel=0&value=1&service=DIG_OUTPUT&me
thod=post&{{CheckTime}}
```

This command will write **digital HIGH** on *channel 0* and *turn* **ON** the LED attached to the Arduino digital pin 3 (also the Grove port D3).

To *write* **digital HIGH** on a selected channel, the *key-value* pair should be written as value=1. Likewise, to write **digital LOW** on a selected channel, the *key-value* pair should be written as value=0.

Figure 8-21. *The Complete Action Fields page*

Modify the following parameters according to your NearBus account's settings:

- **user**: Username of your NearBus account

- **pass**: Password of your NearBus account

- **Device**: Your NearBus device ID

{CheckTime} is a *dummy parameter* that appends the date and time to the message. It can be used to generate a unique message each time, without duplicating the same message, because Twitter rejects duplicate messages to avoid spam.

Here is a sample message generated by our IFTTT recipe.

```
?user=****&pass=****&device=NB100xxx&channel=0&value=1&service=DIG_
OUTPUT&method=post&{{October 10, 2016 at 11:30AM}}
```

15. *Click* the **Create Action** button (Figure 8-21). The **Create and Connect** *page* will appear with the *recipe title. Click* the **Create a Recipe** *button* to create the recipe.

Now you've successfully created your first trigger to turn on the LED. You can repeat these steps to create three more **triggers** to *turn off (after 15 minutes), turn on (after 30 minutes), and turn off (after 45 minutes).*

The four triggers are summarized in Table 8-2.

Table 8-2. *Trigger Commands*

Time (Minutes Past the Hour)	Action	Message Text (Trigger Command)
00	ON	?user=****&pass=****&device=NB102xxx&channel=0&value=1&service=DIG_OUTPUT&method=post&{{CheckTime}}
15	OFF	?user=****&pass=****&device=NB102xxx&channel=0&value=0&service=DIG_OUTPUT&method=post&{{CheckTime}}
30	ON	?user=****&pass=****&device=NB102xxx&channel=0&value=1&service=DIG_OUTPUT&method=post&{{CheckTime}}
45	OFF	?user=****&pass=****&device=NB102xxx&channel=0&value=0&service=DIG_OUTPUT&method=post&{{CheckTime}}

16. After creating the four recipes, you should get a list like the one shown in Figure 8-22.

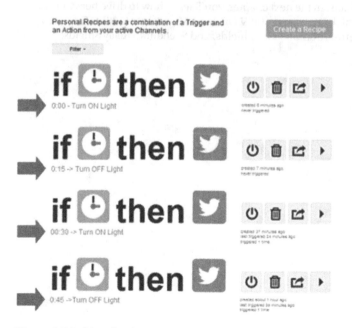

Figure 8-22. *List of recipes*

You've successfully set up the set of IFTTT recipes to trigger ON (digital HIGH) and OFF (digital LOW) NearBus channel 0. The LED will turn on and off every hour at the following times.

- In 0 minutes: Turn ON

- In 15 minutes: Turn OFF

- In 30 minutes: Turn ON

- In 45 minutes: Turn OFF

The triggers will repeat until you turn them off with your IFTTT account or disable your Twitter account.

You can improve this project by replacing the LED and adding an AC-based device with an Arduino digital pin (selected NearBus channel) by connecting a relay driver circuit or PowerSwitch Tail (http://www.powerswitchtail.com/). You can also build this PLC with industry grade enclosures like ArduiBox and an external *power supply* unit like **Meanwell 5V/15W**. This project can be used in industrial environments to turn on and off an industrial exhaust fan every 15 minutes to cool a factory building. You can find many IFTTT receptors to interface with your PLC to control the NearBus cloud connector.

Summary

In this chapter, you learned about the NearBus. Then you built the core hardware setup for a cloud-connectable PLC, mapped your PLC into the cloud using the NearBus cloud connector, created IFTTT recipes to trigger a timer, and finally learned how to control your PLC with Twitter tweets. In the next chapter, you'll learn how to drive heavy loads, like devices that require high DC voltages (48V DC) and AC voltages (120V AC or 240V AC) with single relay boards, Arduino Relay Shields, and Seeeduino Relay Shields.

CHAPTER 9

■ ■ ■

Building a Better PLC

Throughout this book, you've built **Arduino-based PLCs** with **5V** *logic level outputs* and control **5V** *based devices* such as *LEDs, speakers, and fans*. But in industrial environments, you may use loads (devices) that are rated with high voltage (12V DC, 24V DC, 120V AC, or 240V AC) and high current (usually more than 1A).

An Arduino pin can supply **3.3V** or **5V** *voltage levels* with about **20-40mA** of current. These voltage and current levels are useful for building prototypes to test the outputs with Arduino. You can connect various small loads such as *LEDs, speakers, vibrators,* and so forth.

To actuate any device that requires more than 5V of voltage or 40mA of current from the Arduino output signal, a separate *driver circuit* should be connected between the Arduino and the actuator. Some circuit uses *optically-isolated* switches to separate the Arduino output signal from the driver circuit.

Using Relay Boards

A relay-based driver circuit is an ideal solution to drive AC or DC-based loads (devices) rated with high voltage and high power. There are various types of relay driver circuit boards that can be found in the market. The signal input from the microcontroller to the relay board can be *non-optically-isolated or optically-isolated*. It is better to use relay boards that capable with *optically-isolated* signal inputs.

When you work with **AC** or **DC**-based devices, it is safer to use **optically-isolated** relay boards.

Boards with a Single Relay

A **single relay board** consists of a *relay* and a *driver circuit*. Figure 9-1 shows a relay board with a *single relay*, a few electronic components, terminal blocks (**J1**) to connect to the high-voltage or high-power circuit, and male headers (**J2**) to connect to a microcontroller.

© Pradeeka Seneviratne 2017
P. Seneviratne, *Building Arduino PLCs*, DOI 10.1007/978-1-4842-2632-2_9

Figure 9-1. *Single relay board*

Table 9-1 shows the specifications of the **J1** *terminal block.*

Table 9-1. *Specifications of the J1 Terminal Block*

Terminal	Specification
IN	Signal input from the microcontroller
5V	5V from the microcontroller or separate power supply
GND	Ground

Table 9-2 shows the specifications of the **J2** *male header.*

Table 9-2. *Specifications of J2 the Male Header*

Pin	Specification
NO	Connects to the Normally Open pin of the relay.
COM	Connects to the Common pin of the relay.
NC	Connects to the Normally Closed pin of the relay.

You can power this relay with 5V DC and use it directly with Arduino without using any other DC power supply to drive the relay. (*Arduino has a 5V output pin and it can provide enough current to power the relay*).

Let's build a simple project that can be used to turn on and off an **240V AC** *electric light bulb* with an Arduino digital pin.

■ **Note**   If you have a 120V AC power supply in your home, use a 120V AC light bulb with this project. Check your power ratings with the electricity supplier before you build this project.

You'll need the following things to build this project:

- Arduino UNO board

- 5V 1-channel single relay module for Arduino (www.dx.com)

- 240V AC (or 120V AC) light bulb

- Bulb holder

- Two wires (brown for live and blue for neutral), rated for use with 240V AV (or 120V AC)

- Grounded AC power plug

Use the following steps to build the circuit.

1. Using a hook-up wire, connect **Arduino digital pin 13** to the **J2 IN** *header* of the relay board.

2. Connect **Arduino 5V pin** to the **J2 5V** *header* of the **relay board**.

3. Connect **Arduino GND pin** to the **J2 GND** *header* of the **relay board**.

4. Connect the **push button** to **Arduino digital pin 2**.

5. Connect the *brown* (for live) and *blue* (for neutral) wires to the bulb holder and connect the remaining wire ends to a **grounded AC power plug** (Figure 9-2). Use only *live* and *neutral* connections. Then connect the bulb to the holder.

Figure 9-2. Grounded AC power plug

6. Cut the **Live** *wire* of the **electric light bulb** using a wire cutter. You will get two ends. Now connect *one end* to the **J1 NO** *terminal* and the *remaining end* to the **J1 COM** *terminal*.

Now you're ready to upload the sample Arduino sketch to the Arduino board. Open your Arduino IDE and open the sample sketch (Button.ino) by choosing File ➤ Examples ➤ Digital ➤ Button, as shown in Listing 9-1.

Listing 9-1. Electric Light Bulb Test with Button Example (Button.ino)

```
// constants won't change. They're used here to
// set pin numbers:
const int buttonPin = 2;    // the number of the push button pin
const int ledPin = 13;      // the number of the LED pin

// variables will change:
int buttonState = 0;         // variable for reading the push button status

void setup() {
  // initialize the LED pin as an output:
  pinMode(ledPin, OUTPUT);
  // initialize the push button pin as an input:
  pinMode(buttonPin, INPUT);
}

void loop() {
  // read the state of the push button value:
  buttonState = digitalRead(buttonPin);

  // check if the push button is pressed.
  // if it is, the buttonState is HIGH:
  if (buttonState == HIGH) {
    // turn LED on:
    digitalWrite(ledPin, HIGH);
  } else {
    // turn LED off:
    digitalWrite(ledPin, LOW);
  }
}
```

This Arduino sketch can be used with this project because we used the same Arduino pin connections (*digital pin 2 and 13*) for *relay* and *button*. The following Arduino statement is used to assign **Arduino digital pin 13** to the **relay** (**light bulb**).

```
const int ledPin = 13;       // the number of the LED pin
```

Modify the Arduino sketch if you connected the *relay* and *button* to different Arduino digital pins.

1. Verify and upload the sketch to the Arduino board.

2. Remove the Arduino board from the computer and connect it to an external power supply (wall wart).

3. Supply *electricity* (240V AC) to the **light bulb** by connecting the **AC power plug** to an **AC wall socket**.

■ **Note**   You can also use a **Grove Relay** to build this project. Read Chapter 1 or visit http://wiki.seeed.cc/Grove-Relay/ for more information about Grove Relay.

Testing

Now you're ready to test the circuit. When you *press* the **push button**, the light bulb should turn **ON,** and when you *release* it, the light bulb should turn **OFF.**

In industrial environments, you may need to control multiple *high-voltage* and *high-power* loads from Arduino-based PLCs. An **Arduino Relay Shield** is the ideal solution to control multiple loads from the Arduino board; normally you can connect *up to* four loads using a single relay shield.

Boards with Multiple Relays

Like single relay boards, you can use boards with multiple relays to control multiple devices from the Arduino board. Figure 9-3 shows a relay board with two relays that can be used to drive high-voltage and high-current loads.

Figure 9-3. *Relay board with two relays. Image From SparkFun Electronics; photo taken by Juan Peña*

169

Using Relay Shields

Arduino-compatible *relay shields* can be used to replace the *relay boards* because they can be mounted on the Arduino board without using too many wire connections between them. It also helps create more compact projects and is ideal for building Arduino-based PLC projects.

Driving High-Power DC Loads with Relay Shields

In industrial environments, you may need to connect and control *high-power and high-voltage* DC loads with Arduino-based PLCs. The **Arduino 4 Relays Shield,** as shown in Figure 9-4, can be used to connect up to four high-power and high-voltage rated DC devices with Arduino.

With the **Arduino 4 Relays Shield**, you can connect devices rated up to **48V DC**, but it is safer to connect devices rated only to **30V DC/2A**. The shield consists of the following components:

- 4 X 5V relays

- 2 terminal blocks; each terminal block has connections for two devices

- Reset button

- Wire-wrap headers and stackable headers

- 6 x TinkerKit connectors

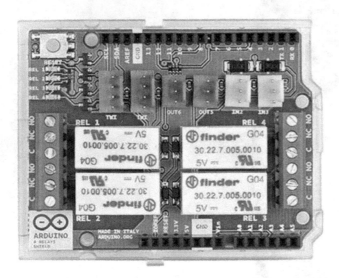

Figure 9-4. Arduino 4 Relays Shield. Image courtesy of arduino.org

The **Arduino 4 Relays Shield** requires 5V and 3.3V power to operate and consumes electricity directly from the **Arduino board**. That means you don't need to provide an extra power supply to the **Arduino 4 Relays shield** with a separate external power supply.

Each relay is internally connected to an **Arduino** *digital pin,* as shown in Table 9-3.

Table 9-3. *Relay and Arduino Digital Pin Assignment*

Relay	Arduino Digital Pin
1	4
2	7
3	8
4	12

The current draw of each *relay coil* is about **35mA**, so with the all relays *on*, it is about **140mA**.

You can find more technical information about the **Arduino 4 Relays Shield** by visiting http://www.arduino.org/products/shields/arduino-4-relays-shield.

Building with an Arduino 4 Relays Shield

To build the following project with an **Arduino 4 Relays Shield**, you need a **light bulb** rated with **12V DC** and a **12V DC power supply** (wall wart). Use the following steps to set up the hardware on the Arduino.

You'll need the following things to build the project.

- Arduino UNO board

- Arduino 4 Relays Shield

- 12V DC light bulb

- Bulb holder

- Two wires (red for positive and black for negative), rated for use with 12V DC

- 12V DC power supply (wall wart)

1. Connect the **Arduino 4 Relays Shield** to the **Arduino UNO** using *wire-wrap* headers.

2. Connect the **push button** to **Arduino** *digital pin 2* with a **10-kilo ohm** *pull-down* resistor.

3. Now cut the *positive* lead of the 12V DC **light bulb** using a wire cutter to make two connection leads.

4. Connect one lead of the **light bulb** to *connector* **C** of the **RELAY 1** *terminal* (Figure 9-5).

5. Connect the remaining lead of the **light bulb** to *connector* **NO** of the **RELAY 1** *terminal*.

NO
NC
C

Figure 9-5. Arduino 4 Relays Shield. Image courtesy of arduino.org

Now you're ready to write and upload the Arduino sketch to the Arduino board.

Modify the Arduino sketch (Button.ino) shown in Listing 9-1 to work with the **RELAY 1** connection and with the **Arduino 4 Relays Shield** because the **RELAY1** is internally connected to the **Arduino** *digital pin 4*.

The following statement shows the required modification for the Arduino sketch:

```
const int ledPin = 4;      // the number of the relay pin
```

Listing 9-2 shows the completed Arduino sketch (ArduinoRelayTestDC.ino) that can be used with the **Arduino 4 Relays Shield.**

Listing 9-2. 12V DC Electric Light Bulb Test with Push Button Example (ArduinoRelayTestDC.ino)

```
// constants won't change. They're used here to
// set pin numbers:
const int buttonPin = 2;    // the number of the push button pin
const int ledPin = 4;       // the number of the relay pin

// variables will change:
int buttonState = 0;        // variable for reading the push button status

void setup() {
  // initialize the LED pin as an output:
  pinMode(ledPin, OUTPUT);
```

172

```
  // initialize the push button pin as an input:
  pinMode(buttonPin, INPUT);
}

void loop() {
  // read the state of the push button value:
  buttonState = digitalRead(buttonPin);

  // check if the push button is pressed.
  // if it is, the buttonState is HIGH:
  if (buttonState == HIGH) {
    // turn LED on:
    digitalWrite(ledPin, HIGH);
  } else {
    // turn LED off:
    digitalWrite(ledPin, LOW);
  }
}
```

1. Verify and upload the Arduino sketch into your Arduino board.

2. Remove the Arduino board from the computer and connect it to an external power supply (wall wart).

3. Supply separate **12V DC** power to the **light bulb** using an external power supply (wall wart).

Testing

Now you're ready to test the hardware setup. When you press the **push button**, the 12V DC light bulb should turn **ON** and when you release it, the light bulb should turn **OFF**.

Driving High-Power AC Loads with Relay Shields

To drive high-power AC loads rated with the high-voltages like **120V AC** and **240V AC**, you can use an Arduino shield that is rated for use with *AC loads*. The **Seeeduino Relay Shield** (Figure 9-6) can be used to drive *high-power AC loads*, for both *120V AC* and *240V AC*. You can also use the Seeeduino Relay Shield to drive *high-power DC loads* with Arduino.

Figure 9-6. Seeeduino Relay Shield v3. Image courtesy of Seeed Development Limited

The **Seeeduino Relay Shield** can be directly stacked on **Arduino UNO** and it can directly draw the power from the Arduino board. Figure 9-7 shows the top view of the **Seeeduino Relay Shield v3.0**.

Figure 9-7. Seeeduino Relay Shield (top view). Image courtesy of Seeed Development Limited

The **Seeeduino Relay shield v3.0** consists of the following components.

- 4 x 5V relays

- 4 x terminal blocks

- 4 x LED indicators

- Wire-wrap headers and stackable headers

You can connect up to four *high-voltage* and *high-power* loads with a Seeeduino Relay Shield and control them individually from the Arduino board.

Each relay is internally connected to an Arduino digital pin, as shown in Table 9-4.

Table 9-4. *Relay and Arduino Digital Pin Assignment*

Relay	Arduino Digital Pin
1	7
2	6
3	5
4	4

Now let's build another PLC with the Seeeduino Relay Shield v3 and an Arduino board. You will need the following components to build the project.

- 10-kilo Ohm resistor

- 240V AC (or 120V AC) light bulb

- Bulb holder

- Two wires (brown for live and blue for neutral), rated for use with 240V AV (or 120V AC)

- Grounded AC power plug

1. Stack the **Seeeduino Relay shield** on top of the **Arduino UNO** board using *wire-wrap* headers.

2. Connect the **push button** to **Arduino** *digital pin 2* with a **10 kilo Ohm** *pull down resistor*.

3. Connect *brown* (for live) and *blue* (for neutral) wires to the bulb holder and connect the remaining wire ends to a **grounded AC power plug**. Use only *live* and *neutral* connections. Then connect the bulb to the holder.

4. Cut the **Live** *wire* of the 120V AC or 240V AC light bulb using a wire cutter. You'll get two wire ends.

5. Use the *terminal block* labeled **CHANNEL 1** of the Seeeduino Relay Shield to connect the light bulb. Connect one end of the **Live** wire to the connector marked **COM** and the remaining end to the connector marked **NO**.

Now you're ready to write and upload the Arduino sketch to the Arduino board.

Modify the Arduino sketch (Button.ino) shown in Listing 9-2 to work with the **RELAY 1** connection and with the **Seeeduino Relay Shield** because the **RELAY1** is internally connected to the **Arduino** *digital pin 7*.

The following statement shows the required modification of the Arduino sketch.

```
const int ledPin =  7;        // the number of the relay pin
```

Listing 9-3 shows the completed Arduino sketch (SeeeduinoRelayTestAC.ino) that can be used with the **Seeeduino Relay Shield**.

Listing 9-3. 240V AC Electric Light Bulb Test with Push Button Example (SeeeduinoRelayTestAC.ino)

```
// constants won't change. They're used here to
// set pin numbers:
const int buttonPin = 2;     // the number of the push button pin
const int ledPin =  7;        // the number of the relay pin

// variables will change:
int buttonState = 0;          // variable for reading the push button status

void setup() {
  // initialize the LED pin as an output:
  pinMode(ledPin, OUTPUT);
  // initialize the push button pin as an input:
  pinMode(buttonPin, INPUT);
}

void loop() {
  // read the state of the push button value:
  buttonState = digitalRead(buttonPin);

  // check if the push button is pressed.
  // if it is, the buttonState is HIGH:
  if (buttonState == HIGH) {
    // turn LED on:
    digitalWrite(ledPin, HIGH);
  } else {
    // turn LED off:
    digitalWrite(ledPin, LOW);
  }
}
```

1. Verify and upload the Arduino sketch to your Arduino board.

2. Remove the Arduino board from the computer and connect it to an external power supply (wall wart).

3. Supply *electricity* (240V AC) to the **light bulb** by connecting the AC power plug to an AC wall socket.

Testing

Now you're ready to test the circuit. When you press the **push button**, the light bulb should turn **ON** and when you release it, the light bulb should turn **OFF**.

Adding More Relay Channels

Seeeduino Relay Shield can be extended with four additional relay channels by connecting another Seeeduino Relay Shield to it. The following steps explain how to connect them.

1. Except for the 5V and 2 GND pins, cut all the wire-wrap headers off the *second* **Seeeduino Relay Shield**.

2. Using hook-up wires, connect the **Seeeduino Relay Shield 1** to the **Seeeduino Relay Shield 2,** as shown in Table 9-5.

Table 9-5. *Wire Connections Between Relay Shield 1 and Relay Shield 2*

Relay Shield 1	Relay Shield 2
Digital pin 8	Digital pin 7
Digital pin 9	Digital pin 6
Digital pin 10	Digital pin 5
Digital pin 11	Digital pin 4

3. Stack the **Seeeduino Relay Shield 2** on top of the **Seeeduino Relay Shield 1** with the remaining three *wire-wrap* headers (5V and 2 GND pins). Use some *standoffs* to properly connect two relay shields together.

Now you can control four relays in the **Seeeduino Relay Shield 2** (*top relay shield*) with the Arduino pins shown in Table 9-6.
Relay Shield 2

Table 9-6. *Pin Assignment with Relays for the Second Relay Shield*

Relay	Terminal	Arduino Digital Pin
1	1	8
2	2	9
3	3	10
4	4	11

As an example, to connect a device to a **Seeeduino Relay Shield 2 - Terminal 1**, the modified Arduino statement would be similar to the following statement.

```
const int ledPin =  8;      // the number of the relay pin
```

Great! You have successfully built a compact relay shield with eight channels by using two Seeeduino Relay Shields and an Arduino board.

Try to connect a *high-voltage device* to any terminal block in the *Seeeduino Relay Shield 2* and control it using the sample Arduino sketch (Button.ino). You do this by modifying the ledPin variable as shown in this statement:

```
const int ledPin =  X     // X = Arduino digital pin to connect to the
relay.
```

Summary

In this chapter, you learned how to modify your PLC to control high-power AC and DC-based devices by using a single relay board (one output) and relay shields (four outputs). You also learned how to add more outputs (eight outputs) by connecting *two* **Seeeduino Relay Shields** together.

A complete Arduino-based PLC can be built with the many techniques we discussed throughout this book. By connecting the following components together, you can build a feature-rich, cloud-connectable PLC.

- ArduiBox enclosure

- 5V external power supply

- 12V DC external power supply (optional)

- Arduino UNO board

- Arduino Ethernet shield

- Relay shield (Arduino 4 Relays Shield or Seeeduino Relay Shield)

Throughout this book you gained knowledge and practical experience on the development of Arduino-based PLC applications with various hardware components, software libraries, communication protocols, passive components like enclosures, and more.

You can modify and enhance these PLC projects by adding various sensors, actuators, power supplies, and optically isolated inputs and outputs in high-voltage environments for AC and DC. As a suggestion, try integrating an LCD (Liquid Crystal Display) into your PLC. It could be useful for presenting the currently executing operation or executed operation to the users. Adding interrupts using a keypad could be also helpful for changing some execution paths manually in some situations.

Index

© Pradeeka Seneviratne 2017
P. Seneviratne, *Building Arduino PLCs*, DOI 10.1007/978-1-4842-2632-2

Get the eBook for only $4.99!

Why limit yourself?

Now you can take the weightless companion with you wherever you go and access your content on your PC, phone, tablet, or reader.

Since you've purchased this print book, we are happy to offer you the eBook for just $4.99.

Convenient and fully searchable, the PDF version enables you to easily find and copy code—or perform examples by quickly toggling between instructions and applications.

To learn more, go to http://www.apress.com/us/shop/companion or contact support@apress.com.

Printed in the United States
By Bookmasters